Handbooks for the Identification of B
Vol. 4 Part 2 (2nd Ed.)

The Carabidae
(ground beetles) of Britain and Ireland
(Second edition)

Martin L. Luff

School of Biology
University of Newcastle upon Tyne

Colour plates prepared by
James Turner

Amgueddfa Cymru – National Museum Wales

© Royal Entomological Society.

1st Edition, Lindroth 1974.

2nd Edition, Luff 2007.

Colour photographs 1-147 in the 2nd Edition
© Amgueddfa Cymru – National Museum Wales, 2007

Published for the Royal Entomological Society
The Mansion House
Bonehill
Chiswell Green Lane
Chiswell Green
St Albans
AL2 3NS
www.royensoc.co.uk

By the Field Studies Council
The Annexe
Preston Montford Lane
Shrewsbury
SY4 1DU
www.field-studies-council.org

ISBN: 978 0 90154 686 9

All rights reserved. No part of this book may be reproduced or translated in any form or by any means, electronically, mechanically, by photocopying or otherwise, without written permission from the copyright holders.

Contents

Acknowledgements	iv	**Key to tribes of Carabinae**	34
Introduction	1	Tribe Carabini	38
Morphology	1	Tribe Cychrini	44
		Tribe Nebriini	45
Biology	4	Tribe Notiophilini	50
Identification	5	Tribe Elaphrini	54
Classification	7	Tribe Loricerini	56
Nomenclature	7	Tribe Scaritini	57
		Tribe Broscini	62
Checklist	10	Tribe Trechini	63
Key to subfamilies of Carabidae	30	Tribe Bembidiini	69
Subfamily Cicindelinae	30	Tribe Pogonini	102
Subfamily Brachininae	32	Tribe Patrobini	103
Subfamily Omophroninae	33	Tribe Pterostichini	105
Subfamily Carabinae	33	Tribes Sphodrini & Platynini	116
		Tribe Zabrini	133
		Tribe Harpalini	148
		Tribe Chlaeniini	177
		Tribe Oodini	179
		Tribe Licinini	180
		Tribe Panagaeini	184
		Tribe Perigonini	185
		Tribe Masoreini	185
		Tribe Lebiini	186
		Tribe Odacanthini	199
		Tribe Dryptini	199
		Tribe Zuphiini	199
		References	201
		Index	203
		Plates	213

Acknowledgements

Many institutions and individuals have helped over the long gestation period of this book. For its original inception, as part of a proposed monograph on British beetles, I am indebted to Mr Andrew Duff, who has continued to provide invaluable advice as editor of the RES Coleoptera Handbooks. For the loan of specimens I wish to thank the Natural History Museum (NHM), London and in particular Mr Max Barclay of the Coleoptera section. Visits to the NHM were partly funded by a grant from the British Entomological and Natural History Society, which is gratefully acknowledged. Mr Roger Booth, also at the NHM gave invaluable advice on nomenclature and provided the initial checklist. Further specimens have kindly been loaned by Professor John Owen and Mr Martin Collier, to whom I am also grateful. For comments on parts of various drafts of the manuscript, I have to thank Mr Max Barclay, Mr Richard Lyskowski, Mr David Nash and Mr Mark Telfer. I thank Mr Peter Hammond for information on *Dyschirius chalceus/nitidus* and the *Badister bullatus* complex. My greatest thanks, however are due to Professor John Owen and Mr Tony Allen, editor of *The Coleopterist*, who have scrutinised the entire text more than once, and have been an invaluable source of guidance and good sense. The colour plates were provided by the staff of the National Museums and Galleries of Wales, Cardiff and my thanks are due especially to Dr Mike Wilson, Mr Brian Levey and Mr Jim Turner for their help and time. I am also grateful to Dr Rebecca Farley of the Field Studies Council for her help in preparation of the manuscript.

It goes without saying that any remaining errors and imperfections in this work remain entirely my responsibility.

Introduction

The Carabidae (ground beetles) are a cosmopolitan family, with an estimated 40,000 species world-wide, about 2700 in Europe and 350 in Britain and Ireland. They are currently considered to be the most developed of the Adephagan families, with their nearest relatives being the aquatic Dytiscidae and Noteridae. The limits of the family have remained constant, in regard to the British and Irish fauna, since the Cicindelidae were included (as subfamily Cicindelinae) by Lindroth (1974). On the larger scale of the European or world fauna however, there have been changes that do not affect the more limited British and Irish lists (see Erwin & Sims, 1984; Erwin, 1985; Lawrence & Newton, 1995).

There have been many changes in classification and nomenclature within the Carabidae since the first Royal Entomological Society Handbook (Lindroth, 1974). Subsequent works on carabid identification elsewhere in Europe include (in English) the fauna of Fennoskandia and Denmark (Lindroth, 1985-86), the Czech and Slovak republics (Hůrka, 1996) and (in German) revised keys to the species of central Europe (Müller-Motzfeld, 2004). A series of illustrated keys to French carabids is also being produced (Forel & Leplat, 2001, 2003, 2005). The present work does not cover the biology, ecology and practical study of ground beetles in any detail, but these are included in the Naturalists' Handbook (Forsythe, 2000). There is also (in Dutch) an exhaustive account of the biology of most of the British and Irish species (Turin, 2000). All aspects of the major genus *Carabus* are included in Turin *et al.* (2003).

Morphology

The distinguishing features of the family are: filiform antennae (at least in the British subfamilies); five-segmented tarsi; hind coxae forming triangular plates that divide the first abdominal segment; hind trochanters lobed, extending part way along the hind margin of the femora. The beetles show a large range in size, from small (1.5 mm) to very large (35 mm) but are fairly uniform in body shape and features. The external morphology of a idealised carabid is shown in Figs 1-2. These should be consulted when reading the following morphological descriptions which refer to the British and Irish groups only and concentrate on features especially useful for identification.

The body is usually rather flattened, especially in species living in crevices in soil such as some species of *Bembidion*, *Pterostichus macer* and *Polistichus* or under bark as in some *Dromius*. A few are more cylindrical (e.g. *Clivina*, *Dyschirius*, Plates 30-32) and live in burrows in the soil. There is a more or less parallel-sided elytral outline, rarely this is rounded (*Omophron*, Plate 4). The body surface is usually mostly glabrous but with scattered setiferous punctures. A few genera are partly or wholly pubescent (e.g. *Ophonus*, *Chlaenius*, Plates 113, 122). Although the majority of British ground beetles are dark-coloured, often black, there are many exceptions to this general rule. Some species are brightly metallic copper, green, purple or have a distinct brassy reflection. In many of these, such as *Harpalus affinis* (Plate 111) the males are both more brightly coloured and shinier than the females. There are distinct patterns of paler spots or other markings on many species, e.g. in *Cicindela* (Plate 1), *Callistus* (Plate 123) and some *Bembidion* and *Badister* (Plates 61, 126). Most species have a fine reticulate microsculpture which is occasionally useful in their identification.

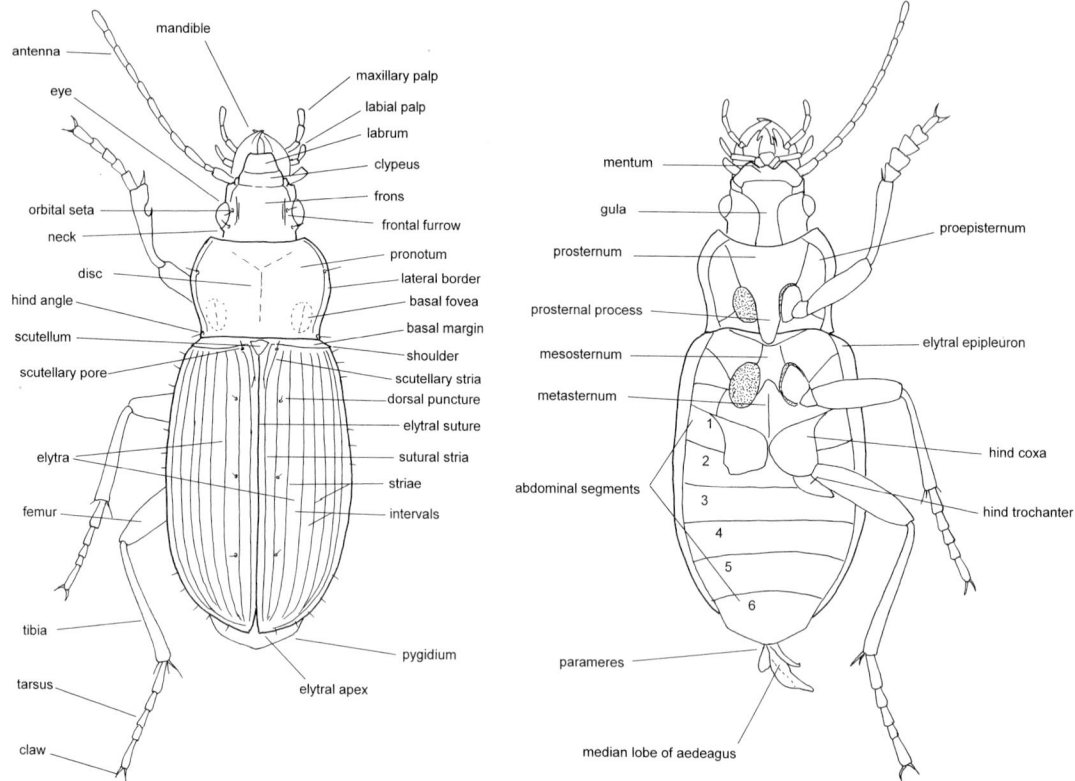

Figure 1. Idealised carabid, dorsal view

Figure 2. Idealised carabid, ventral view

The head has forwardly-protruding mouthparts (prognathous), with a distinct transverse fronto-clypeal suture. In a few genera (e.g. *Cychrus*, *Stomis*, *Odacantha*, Plates 16, 74, 145) it is exceptionally elongate. The anterior margin of the clypeus is toothed in some species (e.g. some *Dyschirius*). On each side of the head are the compound eyes, which are exceptionally large in some diurnal predators (e.g. *Notiophilus*, *Elaphrus*, Plates 25, 27) but reduced in subterranean species (e.g. some *Trechus*, *Aepus*, Plate 36). On the frons inside each eye there may be one or two prominent setae which can be important for identification (Figs 34, 35). The frons may also have semi-circular furrows (as in Trechini) or longitudinal furrows (*Notiophilus*, many *Bembidion*). Behind the eyes there may be a constriction or 'neck'. The underside of the head may have a toothed mentum (e.g. *Agonum*, *Bradycellus*).

The antennae are filiform and 11-segmented. They are situated on the sides of the head outside the mandibles, except in the Cicindelinae, where they are on the dorsal surface behind the mandibles (Plate 1). Most of the antennal segments are usually finely pubescent, but one to four basal segments may be glabrous except for longer apical setae on each segment. In *Loricera* (Plate 29) the antennae have exceptionally long setae. The relative lengths of the basal two or three segments can be useful in identification.

The mouthparts comprise paired mandibles, maxillae and a median labium. The mandibles are toothed internally, and may have a small but taxonomically important setiferous puncture externally. They are slender in many predatory species, broad and triangular in seed-feeders such as some *Harpalus*. Their apices are usually sharp, but are truncate in the snail-feeding Licinini (Fig. 30). The maxillary palpi are four-segmented, the labial palpi three-segmented. The presence and numbers of setae on some of the palpal segments is important taxonomically in species of *Carabus* and in the Zabrini. The apical palpal segments are exceptionally small in the Bembidiini, but enlarged in some other genera such as *Cychrus* and many *Carabus*.

The prothorax is behind the head. The dorsal surface of this is the pronotum, which is usually a flattish rectangular plate, wider than the head with well defined lateral margins. The margins may have a narrow border, the width of which can be taxonomically important, and there may be prominent setae laterally and at the hind angles. In *Odacantha* and *Drypta* (Plates 145, 146) the prothorax is cylindrical, without evident side margins to the pronotum. There are often one or more foveae near the base of the pronotum inside the hind angles. The underneath of the sides (proepisterna) of the prothorax may be punctured, and there may be setae on the hind margin of the prosternal process between the bases of the front legs (e.g. in some *Amara*).

The only part of the mesothorax visible from above is the mesoscutellum (referred to simply as the scutellum), a small triangular plate between the bases of the elytra. In Scaritini and Broscini (Plates 30-33) it is situated in front of the elytral bases on a peduncle behind the pronotum. In *Omophron* the scutellum is hidden by the hind margin of the pronotum.

The remainder of the thorax and most if not all of the abdomen are covered by the wing cases or elytra. These are usually rounded apically but in some groups are truncated as in the Lebiini, etc. (Plates 133-148) exposing the hardened apical abdominal segments called the pygidium. The apical abdominal segments may also protrude in females of species with non-truncated elytra when they have a swollen abdomen full of eggs, or in either sex after preservation in 70% alcohol or other aqueous liquids. The elytral apices are toothed in a few species (e.g. some *Pterostichus* females) and the two elytra may or may not actually meet at the apex. The reflexed undersides (epipleura) of the elytra may be crossed by the elytral margin near the apex ('crossed epipleura' – Fig. 38) as in most of the Pterostichini. The elytra usually have from five to 15 longitudinal grooves or striae, separated by intervals; the striae are often punctured or may be reduced to rows of punctures without grooves. In the Broscini and some *Carabus* striae are absent; in *Carabus*, they are often modified to rows of granules or continuous keels (e.g. Plates 6, 7, 13). The striae and intervals between them are numbered outwards from the elytral mid-line (the suture). Stria 1 is the sutural stria; this may be recurved around the elytral apex as in the Trechini. At the base of the elytra adjacent to the sutural stria there is often an extra short stria called the scutellary stria; this may have a seta and pore (scutellary pore) at its base as in some *Amara*. Other elytral punctures important for identification are:

- dorsal punctures - one to three (rarely five or six) punctures usually on the third interval; they are foveate (i.e. in a depression) in some species;
- humeral punctures – on the outermost interval behind the front elytral angle or shoulder;
- lateral punctures - on the outermost interval at the sides of the elytra; in subterranean species these may bear long setae;
- subapical punctures – in front of the elytral apices usually in the fifth, seventh or eighth intervals.

Beneath the elytra there is often a pair of folded wings, which may be reduced or even lacking in some or all individuals of a species. In the species' descriptions that follow, no distinction is made between fully and partly developed wings: these are simply described as 'present' when the wings are at least as long as the elytra and 'absent' when the wings are shorter then the elytra. Having wings 'present' does not necessarily mean that the species is capable of flight.

The abdomen has six visible ventral segments (except in *Brachinus* with seven or eight – Fig. 6), the first segment being divided by the hind coxae. Punctuation or pubescence on the abdomen may be useful in identification; this can usually be seen from the side in card-mounted specimens. The apical abdominal segments may show secondary sexual characters such as teeth or a ridge, especially in some *Pterostichus*.

The legs are long, usually adapted for running or pushing, occasionally for burrowing (as in *Clivina*, Fig. 15). The hind coxae are immobile and the hind trochanters are elongate, extending along the lateral basal margins of the hind femora. The front tibia usually has an antennal cleaner consisting of a notch and seta on its inner margin (Fig. 14); this is absent in Carabini, Nebriini and some related tribes. The tarsi are five-segmented with a pair of claws which are occasionally finely toothed. In most species the males have broader front (and in some cases also middle) tarsi than the females.

The male genitalia may be needed to confirm identification of some species (especially *Ophonus*) and are almost essential for naming some of the smaller *Badister* species. Elsewhere they may be useful in confirmation of a species identity, but are seldom essential for that purpose. The male genital organ or aedeagus comprises a median lobe and two lateral parameres. The median lobe contains an internal sac which is everted through a subapical opening during mating. The internal sac may contain teeth and other sclerotised structures that are partially visible through the outer wall of the aedeagus. Detailed study of these may require the aedeagus to be cleared and mounted for microscopic examination (see Forsythe, 2000). The internal genital structures have been used as little as possible in this book, as most British and Irish ground beetles can be identified without resource to cleared and mounted male genitalia. The female genitalia may also provide specific characters but have seldom been used in the British fauna except for the *Pterostichus nigrita* complex in which the divided eighth abdominal sternite (usually visible by gently squeezing the abdomen) is used for identification.

Biology

The British and Irish Carabidae are mostly active, terrestrial beetles, which forage on the ground surface, and shelter in litter and under stones, logs etc. The most important habitat characteristic for most species is probably the soil moisture level; many species occur in a rather limited range of moisture conditions, so that there are hygrophilous species such as many *Agonum* and xerophilous species such as most *Harpalus*. The surface activity of most ground beetles means that they can readily be collected by pitfall trapping. Many overwinter in moss, grass tussocks and under loose bark. A few (*Calosoma*, *Dromius*) are truly arboreal, foraging on trees.

Many ground beetles are predatory on small invertebrates; there are specialist feeders on Mollusca (*Cychrus*, Licinini) and on Collembola (*Leistus*, *Notiophilus*, *Loricera*) but others such as most *Carabus* feed on a range of soft-bodied prey including earthworms. The larger *Pterostichus* species are more general scavengers, and will also feed on dead prey as well as some rotting vegetable matter and ripe fruit on the ground. Some Harpalini and Zabrini are phytophagous on seeds or seedlings; *Zabrus* is an occasional pest of cereal seedlings.

Most British and Irish ground beetles have an annual life cycle. They breed either in the spring/summer after adult overwintering, or in the autumn prior to larval overwintering.

Those species that have adult overwintering can be found as inactive adults in hibernacula throughout the winter. Larger species such as *Carabus* may be biennial, or live and breed for two or more seasons so that both larvae and adults overwinter. There may be an obligatory dispersal by flight before breeding, especially in spring-breeders such as many *Bembidion*. The timing and synchronisation of breeding periods has received much attention (see Thiele, 1977) and many species have some form of breeding dormancy or diapause; a well-known example is *Nebria brevicollis* which has an adult summer diapause between spring emergence and autumn breeding.

The immature stages consist of egg, usually three larval instars and pupa. A review of the biology and ecology of these developmental stages is in Luff (2005); only a brief outline is given here. The eggs are usually laid in the soil, in some species protected by a capsule of soil particles moulded by the female. A few genera, such as *Abax*, show parental care of the developing eggs. The larvae of most species forage in the soil; those of the Harpalini and Zabrini are wholly or partly phytophagous but the remainder are primarily predatory. A few genera (e.g. *Carabus*, *Nebria*, *Notiophilus*) have larvae that are active on the soil surface. Larvae of some species of *Ophonus* and *Harpalus* forage on the soil surface for seeds which they then store in subterranean burrows. Larvae of the arboreal *Dromius* species are predatory under bark. Those of *Lebia* and *Brachinus* are ectoparasitic on the pupae of other beetles. Pupation occurs in the larval habitat and the pupal development is usually rather rapid. The newly-emerged adult may then remain in the pupal cavity in the soil for some time before becoming active.

Identification

The prime purpose of this book is to enable reliable identification of the adult Carabidae (ground beetles) that occur in Britain and Ireland. The identification of carabid larvae is not covered in this book but many can be identified using Arndt (1991) or Luff (1993).

Adult carabids to be identified should be surface-dry in order to show fine details of setae, pores and microsculpture. If they are recently removed from liquid preservative, the elytra may need to be raised to show any pale markings, especially in the genus *Bembidion*. A microscope with magnification of up to x40 is needed to see fine details which are best viewed with a diffuse light. Ideally specimens should be identified before being card-mounted, so as to enable examination of the underside. However, underside characters have only been used in the keys where essential, or as confirmation of other features. In some cases these underside features can be seen from the side.

This book is intended to be used systematically, by going through the dichotomous keys, firstly to subfamilies and tribes, then within each of these to genus and species. If the higher taxon of a species is already known then one can start at the keys to a lower level of tribe or genus as appropriate. Users may choose instead to use the colour plates of representatives of each genus and subgenus as a means of jumping straight to their chosen generic or subgeneric key. In this case they should also check the features of their specimen against those listed in the text for that taxon before embarking on what may turn out to be the wrong key.

Notes on each species follow each generic or subgeneric key. These are not complete morphological descriptions but set out the main distinguishing features of that species. They should be used to confirm (or otherwise) the identification made using the preceding keys. Where there are closely-related species that could be mistaken for the subject species, these are cross-referenced in a 'Similar species' paragraph that follows the distinguishing features. These should be used as a further check on the accuracy of identification. The benefit of a reference collection to supplement the keys and for use in conjunction with the descriptions of each species cannot be stressed too highly!

The body length is measured from the front of the mandibles (unless stated otherwise) to the apex of the elytra. In the Lebiini, etc., with truncated elytra and exposed pygidium, the length is measured to the hind margin of the pygidium. The minimum and maximum lengths given for each species are intended to indicate the 'normal' range, within which at least 95% of individuals would fit. Exceptionally small specimens occur in most species; also females are usually larger than males.

Brief details of the typical habitat are then given for each species. This is followed, in roman numerals, by the months in which the species is normally active and foraging or breeding (including both mating and egg-laying), e.g. 'v-viii' for a species typically active from May to August. This does not mean that the species cannot be found outside this period, especially in the winter during adult hibernation.

An approximation of the distribution of each species in Britain and Ireland is included. More detailed accounts, with maps, of the distribution of ground beetles in Britain are in Luff (1998) and on line at www.searchnbn.net. The distribution of carabids in Northern Ireland (with textual information on their occurrence in the Republic) is in Anderson *et al.* (2000) and on line at www.habitas.org.uk/groundbeetles/.

Terms to describe the species' distributions in any part of their range are difficult to define precisely, but are used as follows.
Widespread: may be found in all suitable habitat throughout the range listed.
Local: found in some but not all apparently suitable habitats within the range.
Very local: restricted to a small minority of suitable habitats within the range.
Extremely local: restricted to at most five specific sites within the range.

Finally an attempt is made to give some idea of the abundance of each species where it occurs. This can only be an approximation because of the dynamic nature of any insect population: however it is well known that some species may be extremely local, yet many individuals occur once a population is located, whereas others occur widely, but are seldom found in large numbers. The two terms used to describe abundance are as follows.
Abundant: many individuals (typically more than 10) can easily be found wherever the species occurs. This may be modified by the terms 'very' or 'extremely' to indicate exceptionally high abundances.
Scarce: few individuals (typically less than 10) can be found despite extensive searching or trapping. Again this is modified by 'very' or 'extremely' to indicate exceptionally low numbers.

Classification

There is no general agreement as to the arrangement and status of supra-generic taxa within the Carabidae. In an introduction to the classification of Caraboidea, Ball *et al.* (1998) compared many of the various arrangements that have been advocated. They concluded that points of conflict far outnumbered any generally held areas of agreement. The present list is based on the relatively simple system used by Müller-Motzfeld (2004) which divides the family into only four sub-families, with the great majority of British species in the Carabinae.

This is in contrast to the classification of Lawrence & Newton (1995), who had many subfamilies, and considered the broscines to be a tribe within the subfamily Trechinae. However all other authors either treat them as a separate higher taxon, e.g. family Broscidae alongside family Trechidae (Deuve, 1993), or as a subfamily separate from the trechines e.g. subfamily Broscinae (Erwin, 1985). Detailed studies of the broscines have since supported their separate status (Roig-Juñent, 1998). The trechines are then considered a supertribe Trechitae within the subfamily Psydrinae (Erwin, 1985) or a separate subfamily Trechinae. The present list treats both the broscines and trechines as tribes within the subfamily Carabinae.

A smaller confusion surrounds the placement of the Platynini (often including Sphrodrini). Traditional British and European classifications have included these in the Pterostichini, e.g. Hansen (1996) for Denmark, Turin (2000) for the Netherlands. However Lawrence & Newton (1995) not only agreed with Erwin (1985) in considering the Platynini as a separate tribe, but placed them near the Perigonini, Zuphiini and other 'truncatipennes' groups. The latter placement can be supported on larval characters (Arndt, 1998). Here the Sphodrini and Platynini are retained and are placed immediately after the Pterostichini, as in the traditional arrangement. The tribe Licinini is here considered to include Badistriini, following Kryzhanovskij *et al.* (1995). Subtribe names are not included in the list, although the Harpalini and Platynini, in particular, are often subdivided at this level.

Nomenclature

The nomenclature currently used for the British and Irish Carabidae is largely based on that in Lindroth (1974) and Pope (1977). More recent lists that have updated the nomenclature of some or all of the fauna include Lindroth (1985, Fennoscandia), Lohse & Lucht (1989, central Europe), Turin (1990, 2000, Netherlands), Kryzhanovskij *et al.* (1995, Russia), Hansen (1996, Denmark), Hůrka (1996, Czech & Slovak republics) and Anderson *et al.* (1997, 2000, Ireland). These have now all been superseded by the Palaearctic list edited by Löbl & Smetana (2003). The most recent list of central European Carabidae (Müller-Motzfeld, 2004) does not include synonyms. The checklist in this Handbook is based on that of Löbl & Smetana (2003), with only a few more recent amendments provided by Dr Roger Booth of the Natural History Museum, London. It is also available at www.coleopterist.org.uk.

The treatment of the major genera, *Carabus*, *Bembidion*, *Pterostichus*, *Amara*, *Harpalus* and *Agonum* in the present list has tried to follow the divisions currently generally accepted by most European workers and as used in Löbl & Smetana (2003). This inevitably has resulted in some inconsistencies between the extent of 'splitting' or 'lumping' within these genera.

The limits of *Carabus* are generally agreed, at least for the British fauna, so no 'new' generic names are used. The names and arrangement of subgenera of *Carabus* generally follow Deuve (1994), except that *C. arvensis* is placed in *Eucarabus*, following Turin *et al.* (1993), Kryzhanovskij *et al.* (1995), Hůrka (1996) and Casale & Březina (2003). The classification of *Carabus* in Turin (2000) differs in several ways.

Bembidion, the largest genus of European carabids, is seldom split into separate genera. Only *Ocys* (two British species) and *Cillenus* (a single species) are separated by Kryzhanovskij *et al.* (1995), Hůrka (1996) and Hansen (1996). This Handbook uses these genera, as well as treating *Bracteon* (two species) as a distinct genus, based at least partly on larval features (Luff, 1993). The species related to *andreae* F. have recently been revised by Coulon (2006).

Most European taxonomic works (excepting Hansen, 1996) now consider *Poecilus* to be a separate genus from *Pterostichus*, and this is followed here. The remaining taxa within *Pterostichus* are considered as subgenera, although generic status for some or all may well be justified. Similarly, only *Curtonotus* is separated from *Amara* as a distinct genus, following Kryzhanovskij *et al.* (1995).

The present work follows Telfer (2001b) in considering *Ophonus* as a distinct genus, but does not use the genus *Pseudoophonus*, which remains within *Harpalus* (see Kryzhanovskij *et al.*, 1995). *Agonum* is the only one of the traditionally large genera which is now generally split into a number of genera; these are used here. However *Europhilus* remains as a subgenus of *Agonum*. In addition to the works cited above, the following papers have also been used when constructing the keys in the present book: Sciaky (1987) for *Ophonus* and Serrano & Ortuño (2001) for *Bradycellus*.

Other than changes in synonymies given in the checklist, 13 species have been added to the British and Irish fauna since Pope (1977): the references for these are given in Table 1.

Table 1. Species added to the British list since Pope (1977)

Species	**Basis for addition**	**Literature source**
Asaphidion curtum (Heyden, 1870)	taxonomic split	Speight, Martinez & Luff (1986)
Asaphidion stierlini (Heyden, 1880)	taxonomic split	Speight, Martinez & Luff (1986)
Bembidion caeruleum Audinet-Serville, 1826	immigrant	Telfer (2001a)
Bembidion humerale Sturm, 1825	? rare & overlooked	Crossley & Norris (1975)
Bembidion inustum Jacquelin du Val, 1857	? rare & overlooked	Levey & Pavett (1999)
Pterostichus rhaeticus Heer, 1837/8	taxonomic split	Luff (1990)
Calathus cinctus Motschulsky, 1850	taxonomic split	Anderson & Luff (1994)
Agonum lugens (Duftschmid, 1812)	? rare & overlooked	Anderson (1985)
Harpalus griseus (Panzer, 1797)	immigrant	Owen (1996)
Ophonus subsinuatus Rey, 1886	overlooked museum specimens	Telfer (2001b)
Acupalpus maculatus (Schaum, 1860)	immigrant	Telfer (2003)
Cymindis macularis Mannerheim in Fischer von Waldheim, 1824	? immigrant	Hammond (1982)
Microlestes minutulus (Goeze, 1777)	immigrant	Eversham & Collier (1997)

Thirteen species, either occasional introductions or now considered extinct in Britain and Ireland, have been omitted and are not mentioned in the keys; details are given in Table 2. Further names have been lost from Pope's list due to synonymy or homonomy. This Handbook does not include sub-fossil species no longer occurring in Britain or Ireland.

Table 2. Species not included in the present checklist

Species	Details
Carabus cancellatus Illiger, 1798	Rare introduction.
Carabus convexus Fabricius, 1775	Rare introduction?; recorded from Lancashire in 1836.
Elaphropus quadrisignatus (Duftschmid, 1812)	Rare introduction, probably with ships' ballast; one 19th-Century record from Co. Durham.
Porotachys bisulcatus (Nicolai, 1822)	Temporary introduction, probably with ships' ballast, recorded from Co. Durham in 1863; extinct.
Bembidion callosum Küster, 1847	Rare immigrant?; one specimen found in Surrey in 1851.
Abax parallelus (Duftschmid, 1812)	Temporary introduction, 19th-Century records only, Isles of Scilly; extinct.
Agonum chalconotum Ménétriés, 1832	Possible former resident native, found on banks of R. Clyde, Renfrewshire, to 1914; extinct.
Amara cursitans (Zimmermann, 1832)	Rare immigrant?; two specimens from London in 1950s.
Stenolophus comma (Fabricius, 1775)	Temporary introduction from North America; 19th-Century records only from Renfrewshire.
Lebia marginata (Geoffroy in Fourcroy, 1785)	Rare immigrant or former resident native, two 19th-Century records from Wiltshire and Shropshire; extinct.
Lebia scapularis (Geoffroy in Fourcroy, 1785)	Former resident native, 18th/19th-Century records from Somerset, Middlesex? and Sussex; extinct.
Somotrichus unifasciatus (Dejean, 1831)	Occasional introduction in stored products, not established.
Plochionus pallens (Fabricius, 1775)	Rare introduction in stored products, not established.

Checklist

Family CARABIDAE Latreille, 1802

Subfamily CICINDELINAE Latreille, 1802

CICINDELA Linnaeus, 1758
campestris Linnaeus, 1758 **Green Tiger Beetle**
hybrida Linnaeus, 1758
maritima Latreille & Dejean, 1822 **Dune Tiger Beetle**
sylvatica Linnaeus, 1758 **Wood Tiger Beetle**
silvatica auctt. (misspelling)

CYLINDERA Westwood, 1831
CICINDELA sensu auctt. partim
germanica (Linnaeus, 1758)

Subfamily BRACHININAE Bonelli, 1810

BRACHINUS Weber, 1801
Subgenus BRACHINUS Weber, 1801
crepitans (Linnaeus, 1758) **Bombadier Beetle**
Subgenus BRACHYNIDIUS Reitter, 1919
sclopeta (Fabricius, 1792)

Subfamily OMOPHRONINAE Bonelli, 1810

OMOPHRON Latreille, 1802
SCOLYTUS Fabricius, 1790 non Müller, O.F., 1764
limbatum (Fabricius, 1777)

Subfamily CARABINAE Latreille, 1802

Tribe CARABINI Latreille, 1802

CALOSOMA Weber, 1801
inquisitor (Linnaeus, 1758)
sycophanta (Linnaeus, 1758)

CARABUS Linnaeus, 1758
Subgenus LIMNOCARABUS Géhin, 1876
clatratus Linnaeus, 1761
clathratus auctt. (misspelling)
Subgenus EUCARABUS Géhin, 1885
arvensis Herbst, 1784
ssp. *sylvaticus* Dejean, 1826
silvaticus auctt. (misspelling)
anglicanus Motschulsky, 1865/6
Subgenus CARABUS Linnaeus, 1758
granulatus Linnaeus, 1758
ssp. *granulatus* Linnaeus, 1758

ssp. *hibernicus* Lindroth, 1956
Subgenus MORPHOCARABUS Géhin, 1885
monilis Fabricius, 1792
consitus sensu auctt. non Panzer, 1809
insularis Born, 1908
Subgenus ARCHICARABUS Seidlitz, 1887
nemoralis Müller, O.F., 1764
Subgenus AUTOCARABUS Seidlitz, 1887
auratus Linnaeus, 1761
Subgenus HEMICARABUS Géhin, 1885
nitens Linnaeus, 1758
Subgenus OREOCARABUS Géhin, 1885
glabratus Paykull, 1790
ssp. *lapponicus* Born, 1909
Subgenus MESOCARABUS Thomson, C.G., 1875
problematicus Herbst, 1786
ssp. *feroensis* Lapouge, 1910
ssp. *harcyniae* Sturm, 1815
gallicus Géhin, 1885
catenulatus sensu auctt. non Scopoli, 1763
Subgenus CHAETOCARABUS Thomson, C.G., 1875
intricatus Linnaeus, 1761 **Blue Ground Beetle**
Subgenus MEGODONTUS Solier, 1848
violaceus Linnaeus, 1758 **Violet Ground Beetle**
ssp. *purpurascens* Fabricius, 1787
exasperatus sensu auctt. non Duftschmid, 1812
ssp. *sollicitans* Hartert, 1907
britannicus Born, 1908
browni Deuve, 1999

Tribe CYCHRINI Laporte, 1834

CYCHRUS Fabricius, 1794
caraboides (Linnaeus, 1758)
ssp. *rostratus* (Linnaeus, 1761)

Tribe NEBRIINI Laporte, 1834

LEISTUS Frölich, 1799
Subgenus POGONOPHORUS Latreille, 1802
montanus Stephens, 1827
rufomarginatus (Duftschmid, 1812)
spinibarbis (Fabricius, 1775)
Subgenus LEISTOPHORUS Reitter, 1905
fulvibarbis Dejean, 1826
Subgenus LEISTUS Frölich, 1799
ferrugineus (Linnaeus, 1758)
rufescens sensu auctt. partim non (Fabricius, 1775)
terminatus (Hellwig in Panzer, 1793)
rufescens (Fabricius, 1775) non (Ström, 1768)
praeustus (Fabricius, 1792)

NEBRIA Latreille, 1802
    Subgenus PARANEBRIA Jeannel, 1937
        livida (Linnaeus, 1758)
            lateralis (Fabricius, 1787)
            sabulosa (Fabricius, 1787)
    Subgenus NEBRIA Latreille, 1802
        brevicollis (Fabricius, 1792)
        salina Fairmaire & Laboulbène, 1854
            degenerata Schaufuss, 1862
            iberica Oliveira, 1876
            klinckowstroemi Mjöberg, 1915
    Subgenus BOREONEBRIA Jeannel, 1937
        nivalis (Paykull, 1790)
        rufescens (Ström, 1768)
            gyllenhali (Schönherr, 1806)
            balbii Bonelli, 1810

EURYNEBRIA Ganglbauer, 1891
    complanata (Linnaeus, 1767)

PELOPHILA Dejean, 1821
    borealis (Paykull, 1790)

Tribe NOTIOPHILINI Motschulsky, 1850

NOTIOPHILUS Duméril, 1806
    aesthuans Motschulsky, 1864
        aestuans auctt. (misspelling)
        pusillus Waterhouse, G.R., 1833 non (von Schreber, 1759)
    aquaticus (Linnaeus, 1758)
        pusillus (von Schreber, 1759)
        strigifrons (Baudi, 1864)
        blacki Edwards, J., 1913
    biguttatus (Fabricius, 1779)
    germinyi Fauvel, 1863
        hypocrita sensu auctt. non Putzeys, 1866
    palustris (Duftschmid, 1812)
    quadripunctatus Dejean, 1826
        quadriguttatus Fowler, 1886 (error)
    rufipes Curtis, 1829
    substriatus Waterhouse, G.R., 1833

Tribe ELAPHRINI Latreille, 1802

BLETHISA Bonelli, 1810
    HELOBIUM Leach, 1815
        multipunctata (Linnaeus, 1758)

ELAPHRUS Fabricius, 1775
    Subgenus ELAPHRUS Fabricius, 1775
        cupreus Duftschmid, 1812

lapponicus Gyllenhal, 1810
uliginosus Fabricius, 1792
Subgenus TRICHELAPHRUS Semenov-Tian-Shanskij, 1926
riparius (Linnaeus, 1758)

Tribe LORICERINI Bonelli, 1810

LORICERA Latreille, 1802
LOROCERA auctt. (misspelling)
pilicornis (Fabricius, 1775)
coerulescens sensu auctt. non (Linnaeus, 1758)

Tribe SCARITINI Bonelli, 1810

CLIVINA Latreille, 1802
collaris (Herbst, 1784)
contracta (Geoffroy in Fourcroy, 1785)
fossor (Linnaeus, 1758)

DYSCHIRIUS Bonelli, 1810
Subgenus DYSCHIRIUS Bonelli, 1810
angustatus (Ahrens, 1830)
obscurus (Gyllenhal, 1827)
thoracicus (Rossi, 1790)
arenosus Stephens, 1827
Subgenus DYSCHIRIODES Jeannel, 1941
aeneus (Dejean, 1825)
extensus Putzeys, 1846
elongatulus Dawson, 1856
globosus (Herbst, 1784)
gibbus (Fabricius, 1792)
impunctipennis Dawson, 1854
luedersi Wagner, 1915
aeneus sensu auctt. non (Dejean, 1825)
?*tristis* Stephens, 1827
unicolor sensu auctt. non Motschulsky, 1844
nitidus (Dejean, 1825)
politus (Dejean, 1825)
salinus Schaum, 1843

Tribe BROSCINI Hope, 1838

BROSCUS Panzer, 1813
cephalotes (Linnaeus, 1758)

MISCODERA Eschscholtz, 1830
arctica (Paykull, 1798)

Tribe TRECHINI Bonelli, 1810

PERILEPTUS Schaum, 1860
    BLEMUS sensu Laporte, 1840 non Dejean, 1821
        areolatus (Creutzer, 1799)

AEPUS Leach in Samouelle, 1819
    AEPOPSIS Jeannel, 1922
        marinus (Ström, 1783)
            fulvescens Samouelle, 1819
        robinii (Laboulbène, 1849)
            robini auctt. (misspelling)

TRECHUS Clairville, 1806
    BLEMUS sensu Stephens, 1827 non Dejean, 1821
    Subgenus EPAPHIUS Leach in Samouelle, 1819
        rivularis (Gyllenhal, 1810)
        secalis (Paykull, 1790)
    Subgenus TRECHUS Clairville, 1806
        fulvus Dejean, 1831
            lapidosus (Dawson, 1849)
        obtusus Erichson, 1837
        quadristriatus (Schrank, 1781)
            minutus (Fabricius, 1781)
            rubens sensu Clairville, 1806 non (Fabricius, 1792)
        rubens (Fabricius, 1792)
            paludosus (Gyllenhal, 1810)
        subnotatus Dejean, 1831

THALASSOPHILUS Wollaston, 1854
    TRECHUS sensu Fowler, 1886 partim non Clairville, 1806
        longicornis (Sturm, 1825)

BLEMUS Dejean, 1821
    LASIOTRECHUS Ganglbauer, 1892
        discus (Fabricius, 1792)

TRECHOBLEMUS Ganglbauer, 1891
    micros (Herbst, 1784)

Tribe BEMBIDIINI Stephens, 1827

TACHYS Dejean, 1821
    Subgenus PARATACHYS Casey, 1918
        bistriatus (Duftschmid, 1812)
            pallidulus (Antoine, 1943) non (Ménétriés, 1846)
            pallorus Kopecky, 2003
        micros (Fischer von Waldheim, 1828)
            gregarius Chaudoir, 1846
        obtusiusculus (Jeannel, 1941)
            piceus Edmonds, 1934 non Dalla Torre, 1877
            edmondsi Moore, 1956
    Subgenus TACHYS Dejean, 1821
        scutellaris Stephens, 1828

ELAPHROPUS Motschulsky, 1839
    TACHYS sensu auctt. partim non Dejean, 1821
        parvulus (Dejean, 1831)
        walkerianus (Sharp, 1913)

ASAPHIDION des Gozis, 1886
    TACHYPUS sensu auctt. non Weber, 1801
        curtum (Heyden, 1870)
            flavipes sensu auctt. Brit. partim non (Linnaeus, 1761)
        flavipes (Linnaeus, 1761)
        pallipes (Duftschmid, 1812)
        stierlini (Heyden, 1880)
            flavipes sensu auctt. Brit. partim non (Linnaeus, 1761)

OCYS Stephens, 1828
    BEMBIDION sensu auctt. partim non Latreille, 1802
        harpaloides (Audinet-Serville, 1821)
            rufescens (Guérin-Méneville, 1823)
        quinquestriatus (Gyllenhal, 1810)

CILLENUS Leach in Samouelle, 1819
    BEMBIDION sensu auctt. partim non Latreille, 1802
        lateralis Samouelle, 1819

BRACTEON Bedel, 1879
    BEMBIDION sensu auctt. partim non Latreille, 1802
    CHRYSOBRACTEON Netolitzky, 1914
        argenteolum (Ahrens, 1812)
        litorale (Olivier, 1790)
            littorale (Olivier, 1791)
        paludosum (Panzer, 1794)

BEMBIDION Latreille, 1802
    BEMBIDIUM auctt. (misspelling)
    Subgenus NEJA Motschulsky, 1864
        nigricorne Gyllenhal, 1827
    Subgenus METALLINA Motschulsky, 1850
        lampros (Herbst, 1784)
            celere (Fabricius, 1792)
        properans (Stephens, 1828)
            velox Erichson, 1837 non (Linnaeus, 1761)
            coeruleotinctum Reitter, 1908
            cyaneotinctum Sharp, 1913
            caeruleipenne Saunders, 1936
    Subgenus PRINCIDIUM Motschulsky, 1864
        punctulatum Drapiez, 1821
    Subgenus ACTEDIUM Motschulsky, 1864
        pallidipenne (Illiger, 1802)
            ruficolle (Illiger, 1801) non (Panzer, 1796)
    Subgenus TESTEDIUM Motschulsky, 1864
        bipunctatum (Linnaeus, 1761)

Subgenus EUPETEDROMUS Netolitzky, 1911
    dentellum (Thunberg, 1787)
        flammulatum (Clairville, 1806)
Subgenus NOTAPHUS Dejean, 1821
    obliquum Sturm, 1825
    semipunctatum (Donovan, 1806)
        adustum Schaum, 1860
    varium (Olivier, 1795)
        ustulatum sensu Sturm, 1825 non (Linnaeus, 1758)
        nebulosum (Stephens, 1828)
Subgenus NOTAPHEMPHANES Netolitzky, 1920
    NOTHAPHEMPHANES auctt. (misspelling)
    ephippium (Marsham, 1802)
Subgenus PLATAPHUS Motschulsky, 1864
    prasinum (Duftschmid, 1812)
Subgenus TRICHOPLATAPHUS Netolitzky, 1914
    BLEPHAROPLATAPHUS Netolitzky, 1920
    virens Gyllenhal, 1827
Subgenus BEMBIDIONETOLITZKYA Strand, E., 1929
    DANIELA Netolitzky, 1910 non Koch, 1891
    atrocaeruleum (Stephens, 1828)
        atrocoeruleum auctt. (misspelling)
    caeruleum Audinet-Serville, 1826
        coeruleum auctt. (misspelling)
    geniculatum Heer, 1837/8
        redtenbacheri Daniel, K., 1902
    tibiale (Duftschmid, 1812)
Subgenus OCYDROMUS Clairville, 1806
    PERYPHUS Dejean, 1821
    bruxellense Wesmael, 1835
        rupestre sensu auctt. non (Linnaeus, 1767)
    bualei Jacquelin du Val, 1852
        andreae sensu auctt. non (Fabricius, 1787)
        cruciatum sensu auctt. non Dejean, 1831
        ssp. *anglicanum* Sharp, 1869
        ssp. *polonicum* Müller, 1930
    decorum (Zenker in Panzer, 1800)
    deletum Audinet-Serville, 1821
        nitidulum (Marsham, 1802) non (Schrank, 1781)
        dalmatinum var. *latinum* sensu MacKechnie-Jarvis, 1932 non Netolitzky, 1911
    femoratum Sturm, 1825
    fluviatile Dejean, 1831
    lunatum (Duftschmid, 1812)
    maritimum (Stephens, 1835)
        concinnum sensu auctt. non (Stephens, 1828)
        dorsuarium Bedel, 1879
    monticola Sturm, 1825
    saxatile Gyllenhal, 1827
        vectense Fowler, 1886

    stephensii Crotch, 1866
      stephensi auctt. (misspelling)
      affine (Stephens, 1835) non Say, 1825
    testaceum (Duftschmid, 1812)
    tetracolum Say, 1825
      ustulatum sensu auctt. non (Linnaeus, 1758)
      littorale sensu auctt. non (Olivier, 1791)
  Subgenus NEPHA Motschulsky, 1864
    illigeri Netolitzky, 1914
      quadriguttatum sensu auctt. Brit. non (Illiger, 1798) nec (Fabricius, 1775)
      tetragrammum auctt. non Chaudoir, 1846
      genei sensu auctt. Brit. non Küster, 1847
  Subgenus SINECHOSTICTUS Motschulsky, 1864
    SYNECHOSTICTUS auctt. (misspelling)
    stomoides Dejean, 1831
      atroviolaceum sensu auctt. non Dufour, 1820
  Subgenus PSEUDOLIMNAEUM Kraatz, 1888
    inustum Jacquelin du Val, 1857
  Subgenus LYMNAEUM Stephens, 1828
    nigropiceum (Marsham, 1802)
  Subgenus SEMICAMPA Netolitzky, 1910
    gilvipes Sturm, 1825
    schuppelii Dejean, 1831
      schueppeli auctt. (misspelling)
  Subgenus DIPLOCAMPA Bedel, 1896
    assimile Gyllenhal, 1810
    clarkii (Dawson, 1849)
      clarki auctt. (misspelling)
    fumigatum (Duftschmid, 1812)
  Subgenus EMPHANES Motschulsky, 1850
    minimum (Fabricius, 1792)
    normannum Dejean, 1831
  Subgenus BEMBIDION Latreille, 1802
    LOPHA Dejean, 1821 non Bolten, 1798
    humerale Sturm, 1825
    quadrimaculatum (Linnaeus, 1761)
      quadriguttatum (Fabricius, 1775)
    quadripustulatum Audinet-Serville, 1821
      quadriguttatum sensu (Olivier, 1795) non (Fabricius, 1775)
      antiquorum Crotch, 1871
  Subgenus TREPANEDORIS Netolitzky, 1918
    doris (Panzer, 1796)
  Subgenus TREPANES Motschulsky, 1864
    LEJA Dejean, 1821
    articulatum (Panzer, 1795)
    octomaculatum (Goeze, 1777)
      sturmii (Panzer, 1804)
      sturmi auctt. (misspelling)

Subgenus PHYLA Motschulsky, 1844
    PHILA Motschulsky, 1846 (*nom. emend.*)
    obtusum Audinet-Serville, 1821
Subgenus PHILOCHTHUS Stephens, 1828
    PHILOCTHUS auctt. (misspelling)
    aeneum Germar, 1824
    biguttatum (Fabricius, 1779)
    guttula (Fabricius, 1792)
    iricolor Bedel, 1879
        riparium sensu Fowler, 1886 partim non (Olivier, 1795)
    lunulatum (Geoffroy in Fourcroy, 1785)
        riparium (Olivier, 1795)
    mannerheimii Sahlberg, C.R., 1827
        mannerheimi auctt. (misspelling)
        haemorrhoum sensu auctt. non (Stephens, 1828)
    unicolor Chaudoir, 1850

Tribe POGONINI Laporte, 1834

POGONUS Dejean, 1821
    chalceus (Marsham, 1802)
    littoralis (Duftschmid, 1812)
        litoralis auctt. (misspelling)
    luridipennis (Germar, 1822)

Tribe PATROBINI Kirby, 1837

PATROBUS Dejean, 1821
    assimilis Chaudoir, 1844
        clavipes Thomson, C.G., 1857
    atrorufus (Ström, 1768)
        excavatus (Paykull, 1790)
        rufipes sensu (Duftschmid, 1812) non (Fabricius, 1792)
    septentrionis Dejean, 1828

Tribe PTEROSTICHINI Bonelli, 1810

STOMIS Clairville, 1806
    pumicatus (Panzer, 1795)

POECILUS Bonelli, 1810
    PTEROSTICHUS sensu auctt. partim non Bonelli, 1810
    FERONIA Latreille, 1817
    cupreus (Linnaeus, 1758)
        affinis (Sturm, 1824)
        erythropus (Dejean, 1828)
        dinniki (Lutshnik, 1912)
        caesica Donisthorpe, 1931
    kugelanni (Panzer, 1797)
        dimidiatus (Olivier, 1795) non (Rossi, 1790)
    lepidus (Leske, 1785)
    versicolor (Sturm, 1824)
        caerulescens sensu auctt. non (Linnaeus, 1758)

PTEROSTICHUS Bonelli, 1810
    Subgenus PTEROSTICHUS Bonelli, 1810
        cristatus (Dufour, 1820)
            ssp. *parumpunctatus* Germar, 1824
    Subgenus STEROPUS Stephens, 1828
        aethiops (Panzer, 1796)
        madidus (Fabricius, 1775)
            concinnus (Sturm, 1818)
    Subgenus PEDIUS Motschulsky, 1850
        longicollis (Duftschmid, 1812)
            inaequalis (Marsham, 1802) non (Panzer, 1795)
        ochraceus (Sturm, 1824)
    Subgenus LYPEROSOMUS Motschulsky, 1850
        aterrimus (Herbst, 1784)
    Subgenus ADELOSIA Stephens, 1835
        macer (Marsham, 1802)
            picimanus (Duftschmid, 1812)
    Subgenus PLATYSMA Bonelli, 1810
        niger (Schaller, 1783)
            scotus (Jeannel, 1942)
    Subgenus BOTHRIOPTERUS Chaudoir, 1838
        adstrictus Eschscholtz, 1823
            orinomum (Stephens, 1828)
            vitreus (Dejean, 1828)
        oblongopunctatus (Fabricius, 1787)
        quadrifoveolatus Letzner, 1852
            angustatus (Duftchmid, 1812) non (Fabricius, 1787)
    Subgenus OMASEUS Stephens, 1828
        melanarius (Illiger, 1798)
            vulgaris sensu auctt. non (Linnaeus, 1758)
    Subgenus PSEUDOMASEUS Chaudoir, 1838
        anthracinus (Panzer, 1795)
        gracilis (Dejean, 1828)
        minor (Gyllenhal, 1827)
        nigrita (Paykull, 1790)
        rhaeticus Heer, 1837/8
            nigrita sensu auctt. partim non (Paykull, 1790)
    Subgenus LAGARUS Chaudoir, 1838
        vernalis (Panzer, 1795)
            crenatus (Duftschmid, 1812) non (Gmelin in Linnaeus, 1790)
    Subgenus ARGUTOR Stephens, 1828
        diligens (Sturm, 1824)
            strenuus sensu Dawson, 1854 non (Panzer, 1796)
        strenuus (Panzer, 1796)
            erythropus (Marsham, 1802)

ABAX Bonelli, 1810
        parallelepipedus (Piller & Mitterpacher, 1783)
            parallelopipedus auctt. (misspelling)
            ater (Villers, 1790)
            striola (Fabricius, 1792)

Tribe SPHODRINI Laporte, 1834

 PLATYDERUS Stephens, 1828
  PLATYDERES Stephens, 1827
   depressus (Audinet-Serville, 1821)
    ruficollis (Marsham, 1802) non (Fabricius, 1787)

 SYNUCHUS Gyllenhal, 1810
  TAPHRIA Latreille, 1819
  ODONTONYX Stephens, 1827
   vivalis (Illiger, 1798)
    nivalis (Panzer, 1797) non (Paykull, 1790)

 CALATHUS Bonelli, 1810
  Subgenus AMPHIGYNUS Haliday, 1841
   rotundicollis Dejean, 1828
    piceus sensu (Marsham, 1802) non (Linnaeus, 1758)
  Subgenus CALATHUS Bonelli, 1810
   ambiguus (Paykull, 1790)
    fuscus (Fabricius, 1792)
   cinctus Motschulsky, 1850
    mollis sensu auctt. partim non Marsham, 1802
    erythroderus Gemminger & Harold, 1868
   erratus (Sahlberg, C.R., 1827)
    fulvipes sensu (Gyllenhal, 1810) non (Fabricius, 1792)
    flavipes sensu (Duftschmid, 1812) non (Geoffroy in Fourcroy, 1785)
   fuscipes (Goeze, 1777)
    cisteloides (Panzer, 1793)
   melanocephalus (Linnaeus, 1758)
    ochropterus (Duftschmid, 1812)
   nubigena Haliday, 1838
   micropterus (Duftschmid, 1812)
   mollis (Marsham, 1802)
    ochropterus sensu auctt. non (Duftschmid, 1812)

 SPHODRUS Clairville, 1806
  leucophthalmus (Linnaeus, 1758)
   leucopthalmus Fowler, 1886 (misspelling)
   planus (Fabricius, 1792)

 LAEMOSTENUS Bonelli, 1810
  LAEMOSTHENES Schaufuss, 1865
  Subgenus LAEMOSTENUS Bonelli, 1810
   complanatus (Dejean, 1828)
  Subgenus PRISTONYCHUS Dejean, 1828
   terricola (Herbst, 1784)
    subcyaneus (Illiger, 1801)

Tribe PLATYNINI Bonelli, 1810

OLISTHOPUS Dejean, 1828
  ODONTONYX sensu auctt. non Stephens, 1827
    rotundatus (Paykull, 1790)
      rotundicollis (Marsham, 1802)

OXYPSELAPHUS Chaudoir, 1843
  AGONUM sensu auctt. partim non Bonelli, 1810
  ANCHUS LeConte, 1854
    obscurus (Herbst, 1784)
      ?*obscurus* (Müller, O.F., 1776)
      oblongus (Fabricius, 1792)

PARANCHUS Lindroth, 1974
  AGONUM sensu auctt. partim non Bonelli, 1810
    albipes (Fabricius, 1796)
      ruficornis (Goeze, 1777) non (De Geer, 1774)
      oblongus (Fabricius, 1792) non (Fabricius, 1792)
      pallipes (Fabricius, 1801) non (Fabricius, 1787)

ANCHOMENUS Bonelli, 1810
  AGONUM sensu auctt. partim non Bonelli, 1810
    dorsalis (Pontoppidan, 1763)
      prasinus (Thunberg, 1784)

PLATYNUS Bonelli, 1810
  AGONUM sensu auctt. partim non Bonelli, 1810
    assimilis (Paykull, 1790)
      ?*junceus* (Scopoli, 1763)
      angusticollis (Fabricius, 1801)

BATENUS Motschulsky, 1864
    livens (Gyllenhal, 1810)

SERICODA Kirby, 1837
  AGONUM sensu auctt. partim non Bonelli, 1810
    quadripunctata (De Geer, 1774)

AGONUM Bonelli, 1810
  Subgenus EUROPHILUS Chaudoir, 1859
    fuliginosum (Panzer, 1809)
    gracile Sturm, 1824
    micans Nicolai, 1822
    piceum (Linnaeus, 1758)
    scitulum Dejean, 1828
    thoreyi Dejean, 1828
      pelidnum sensu (Paykull, 1792) non (Herbst, 1784)
      puellum Dejean, 1828

Subgenus AGONUM Bonelli, 1810
    emarginatum (Gyllenhal, 1827)
        afrum (Duftschmid, 1812) non (Thunberg, 1784)
        moestum sensu auctt. non (Duftschmid, 1812) nec (Gmelin in Linnaeus, 1790)
    ericeti (Panzer, 1809)
    gracilipes (Duftschmid, 1812)
        elongatum Dejean, 1828
    lugens (Duftschmid, 1812)
    marginatum (Linnaeus, 1758)
    muelleri (Herbst, 1784)
        parumpunctatum (Fabricius, 1792)
        chalybeum Sturm, 1824
    nigrum Dejean, 1828
        atratum sensu auctt. non (Duftschmid, 1812)
        dahli (Preudhomme de Borre, 1879)
    sexpunctatum (Linnaeus, 1758)
    versutum Sturm, 1824
    viduum (Panzer, 1796)
        moestum sensu Fowler & Donisthorpe, 1913 non (Duftschmid, 1812)

Tribe ZABRINI Bonelli, 1810

ZABRUS Clairville, 1806
    tenebrioides (Goeze, 1777)
        gibbus (Fabricius, 1794)

AMARA Bonelli, 1810
    Subgenus ZEZEA Csiki, 1929
        TRIAENA LeConte, 1847 non Hübner, 1818
    plebeja (Gyllenhal, 1810)
        plebeia auctt. (misspelling)
    strenua Zimmermann, 1832
        vectensis Dawson, 1849
    Subgenus AMARA Bonelli, 1810
    aenea (De Geer, 1774)
        trivialis (Gyllenhal, 1810)
    anthobia Villa & Villa, 1833
    communis (Panzer, 1797)
        vulgaris sensu Dawson, 1854 non (Linnaeus, 1758)
    convexior Stephens, 1828
        continua Thomson, C.G., 1873
    curta Dejean, 1828
    eurynota (Panzer, 1796)
        acuminata (Paykull, 1798)
    famelica Zimmermann, 1832
    familiaris (Duftschmid, 1812)
    lucida (Duftschmid, 1812)
    lunicollis Schiödte, 1837
        vulgaris sensu auctt. non (Linnaeus, 1758)

   montivaga Sturm, 1825
   nitida Sturm, 1825
   ovata (Fabricius, 1792)
    obsoleta sensu Dejean, 1828 non (Duftschmid, 1812)
    adamantina Kolenati, 1845
   similata (Gyllenhal, 1810)
    obsoleta (Duftschmid, 1812)
   spreta Dejean, 1831
   tibialis (Paykull, 1798)
  Subgenus CELIA Zimmermann, 1832
   bifrons (Gyllenhal, 1810)
    livida sensu Schiödte, 1841 non (Fabricius, 1792)
   fusca Dejean, 1828
    complanata Dejean, 1828
   infima (Duftschmid, 1812)
   praetermissa (Sahlberg, C.R., 1827)
    rufocincta (Sahlberg, C.R., 1827)
  Subgenus PARACELIA Bedel, 1899
   quenseli (Schönherr, 1806)
    quenselii auctt.
  Subgenus BRADYTUS Stephens, 1827
   apricaria (Paykull, 1790)
   consularis (Duftschmid, 1812)
   fulva (Müller, O.F., 1776)
  Subgenus PERCOSIA Zimmermann, 1832
   equestris (Duftschmid, 1812)
    patricia (Duftschmid, 1812)

CURTONOTUS Stephens, 1827
 CYRTONOTUS auctt. (misspelling)
 AMARA sensu auctt. partim non Bonelli, 1810
  alpinus (Paykull, 1790)
  aulicus (Panzer, 1796)
   spinipes sensu Schiödte, 1841 non (Linnaeus, 1758)
  convexiusculus (Marsham, 1802)

Tribe HARPALINI Bonelli, 1810

HARPALUS Latreille, 1802
 Subgenus PSEUDOOPHONUS Motschulsky, 1844
  PSEUDOPHONUS auctt. (misspelling)
  PARDILEUS des Gozis, 1882
  calceatus (Duftschmid, 1812)
  griseus (Panzer, 1797)
  rufipes (De Geer, 1774)
   ruficornis (Fabricius, 1775)
   pubescens (Müller, O.F., 1776)
 Subgenus HARPALUS Latreille, 1802
  HAPLOHARPALUS Schauberger, 1926
  affinis (Schrank, 1781)
   aeneus (Fabricius, 1775) non (De Geer, 1774)
  anxius (Duftschmid, 1812)

  attenuatus Stephens, 1828
    consentaneus Dejean, 1829
  cupreus Dejean, 1829
  dimidiatus (Rossi, 1790)
    caspius sensu auctt. Brit. non (von Steven, 1806)
  froelichii Sturm, 1818
    froelichi auctt. (misspelling)
  honestus (Duftschmid, 1812)
    ignavus (Duftschmid, 1812)
  laevipes Zetterstedt, 1828
    seriepunctatus sensu Gyllenhal, 1827 non Sturm, 1818
    quadripunctatus Dejean, 1829
    montivagus Reitter, 1900
  latus (Linnaeus, 1758)
    erythrocephalus (Fabricius, 1787)
    metallescens Rye, 1874
  neglectus Audinet-Serville, 1821
  pumilus Sturm, 1818
    vernalis sensu (Duftschmid, 1812) non (Panzer, 1796)
    picipennis sensu auctt. Brit. non (Duftschmid, 1812)
    funestus Audinet-Serville, 1821
  rubripes (Duftschmid, 1812)
    sobrinus Dejean, 1829
  rufipalpis Sturm, 1818
    rufitarsis (Duftschmid, 1812) non (Illiger, 1801)
    ignavus sensu auctt. Brit. non (Duftschmid, 1812)
  serripes (Quensel in Schönherr, 1806)
  servus (Duftschmid, 1812)
  smaragdinus (Duftschmid, 1812)
    discoideus Erichson, 1837
  tardus (Panzer, 1796)
    rufimanus (Marsham, 1802)
    luteicornis sensu auctt. Brit. non (Duftschmid, 1812)
 Subgenus CRYPTOPHONUS Brandmayr & Zetto Brandmayr, 1982
  melancholicus Dejean, 1829
  tenebrosus Dejean, 1829
    ssp. *centralis* Schauberger, 1929

OPHONUS Dejean, 1821
 HARPALUS sensu auctt. partim non Latreille, 1802
 Subgenus OPHONUS Dejean, 1821
  ardosiacus Lutshnik, 1922
    obscurus sensu (Sturm, 1818) non (Fabricius, 1792)
    diffinis sensu Joy, 1932 non (Dejean, 1829)
    rotundicollis sensu auctt. Brit. non Kolenati, 1845
  azureus (Fabricius, 1775)
    subquadratus sensu auctt. Brit. non (Dejean, 1829)
  sabulicola (Panzer, 1796)
  stictus Stephens, 1828
    obscurus (Fabricius, 1792) non (Müller, O.F., 1776)
    monticola (Dejean, 1829)

Subgenus METOPHONUS Bedel, 1895
   cordatus (Duftschmid, 1812)
   laticollis Mannerheim, 1825
      punctatulus (Duftschmid, 1812) non (Fabricius, 1792)
      nitidulus sensu Stephens, 1828 non (Schrank, 1781)
   melletii (Heer, 1837/8)
      melleti auctt. (misspelling)
      rectangulus Thomson, C.G., 1870
      championi Sharp, 1912
      rupicoloides Sharp, 1912
      brevicollis sensu Jeannel, 1942 ? (Audinet-Serville, 1821)
   parallelus (Dejean, 1829)
      melleti sensu Jeannel, 1942 non (Heer, 1837/8)
      zigzag sensu auctt. Brit. non Costa, 1882
   puncticeps Stephens, 1828
      rectangulus sensu Sharp, 1912 non Thomson, C.G., 1870
      angusticollis Müller, J., 1921
   puncticollis (Paykull, 1798)
   rufibarbis (Fabricius, 1792)
      brevicollis sensu auctt. partim non (Audinet-Serville, 1821)
      subpunctatus Stephens, 1828
      seladon (Schauberger, 1926)
   rupicola (Sturm, 1818)
   schaubergerianus (Puel, 1937)
      rufibarbis sensu auctt. non (Fabricius, 1792)
      brevicollis sensu auctt. partim non (Audinet-Serville, 1821)
   subsinuatus Rey, 1886

ANISODACTYLUS Dejean, 1829
   binotatus (Fabricius, 1787)
      spurcaticornis Dejean, 1829
   nemorivagus (Duftschmid, 1812)
      atricornis (Stephens, 1835)
   poeciloides (Stephens, 1828)
      pseudoaeneus sensu auctt. non Dejean, 1829

DIACHROMUS Erichson, 1837
   germanus (Linnaeus, 1758)

SCYBALICUS Schaum, 1862
   oblongiusculus (Dejean, 1829)

DICHEIROTRICHUS Jacquelin du Val, 1855
  DICHIROTRICHUS auctt. (misspelling)
   gustavii Crotch, 1871
      gustavi auctt. (misspelling)
   obsoletus (Dejean, 1829)
      pubescens (Paykull, 1790) non (Müller, O.F., 1776)

TRICHOCELLUS Ganglbauer, 1892
   BRADYCELLUS sensu Fowler, 1886 partim non Erichson, 1837

cognatus (Gyllenhal, 1827)
placidus (Gyllenhal, 1827)

BRADYCELLUS Erichson, 1837
caucasicus (Chaudoir, 1846)
collaris (Paykull, 1798) non (Herbst, 1784)
csikii Laczó, 1912
distinctus (Dejean, 1829)
harpalinus (Audinet-Serville, 1821)
ruficollis (Stephens, 1828)
similis (Dejean, 1829)
sharpi Joy, 1912
distinctus sensu Fowler, 1886 non (Dejean, 1829)
verbasci (Duftschmid, 1812)

STENOLOPHUS Dejean, 1821
AGONODERUS Dejean, 1829
mixtus (Herbst, 1784)
vespertinus (Panzer, 1796)
ziegleri (Panzer, 1809)
skrimshiranus Stephens, 1828
teutonus (Schrank, 1781)
vaporariorum sensu (Fabricius, 1775) non (Linnaeus, 1758)
anglicus Schiödte, 1861-63

ACUPALPUS Latreille, 1829
brunnipes (Sturm, 1825)
brunneipes auctt. (misspelling)
dubius Schilsky, 1888
luridus sensu auctt. non Dejean, 1829
luteatus sensu Joy, 1932 non (Duftschmid, 1812)
elegans (Dejean, 1829)
exiguus Dejean, 1829
flavicollis (Sturm, 1825)
luridus Dejean, 1829
maculatus (Schaum, 1860)
meridianus (Linnaeus, 1761)
parvulus (Sturm, 1825)
dorsalis (Fabricius, 1787) non (Pontoppidan, 1763)
derelictus (Dawson, 1854)

ANTHRACUS Motschulsky, 1850
ACUPALPUS sensu auctt. partim non Latreille, 1829
consputus (Duftschmid, 1812)

Tribe CHLAENIINI Laporte, 1834

CHLAENIUS Bonelli, 1810
CHLAENIELLUS Reitter, 1908
nigricornis (Fabricius, 1787)
melanocornis Dejean, 1826

        nitidulus (Schrank, 1781)
           schrankii (Duftschmid, 1812)
        tristis (Schaller, 1783)
           holosericeus (Fabricius, 1787)
        vestitus (Paykull, 1790)

CALLISTUS Bonelli, 1810
        lunatus (Fabricius, 1775)

Tribe OODINI LaFerte-Senectere, 1851

OODES Bonelli, 1810
        helopioides (Fabricius, 1792)

Tribe LICININI Bonelli, 1810

LICINUS Latreille, 1802
        depressus (Paykull, 1790)
        punctatulus (Fabricius, 1792)
           punctulatus Kloet & Hincks, 1945 (error)
           silphoides sensu (Fabricius, 1792) non (Rossi, 1790)

BADISTER Clairville, 1806
    Subgenus BADISTER Clairville, 1806
        bullatus (Schrank, 1798)
           bipustulatus (Fabricius, 1792) non (Fabricius, 1775)
        meridionalis Puel, 1925
           kineli Makolski, 1952
        unipustulatus Bonelli, 1813
    Subgenus TRIMORPHUS Stephens, 1828
        sodalis (Duftschmid, 1812)
           humeralis Bonelli, 1813
    Subgenus BAUDIA Ragusa, 1884
        collaris Motschulsky, 1844
           anomalus (Perris, 1866)
           striatulus Hansen, 1944
        dilatatus Chaudoir, 1837
        peltatus (Panzer, 1797)

Tribe PANAGAEINI Bonelli, 1810

PANAGAEUS Latreille, 1802
        bipustulatus (Fabricius, 1775)
           quadripustulatus Sturm, 1815
        cruxmajor (Linnaeus, 1758)

Tribe PERIGONINI Horn, 1881

PERIGONA Laporte, 1835
        nigriceps (Dejean, 1831)

Tribe MASOREINI Chaudoir, 1870

MASOREUS Dejean, 1821
    wetterhallii (Gyllenhal, 1813)
        wetterhali auctt. (misspelling)
        laticollis (Sturm, 1825)

Tribe LEBIINI Bonelli, 1810

LEBIA Latreille, 1802
    Subgenus **LAMPRIAS** Bonelli, 1810
        chlorocephala (Hoffmann, J., 1803)
            chrysocephala (Motschulsky, 1864)
        cyanocephala (Linnaeus, 1758)
    Subgenus **LEBIA** Latreille, 1802
        cruxminor (Linnaeus, 1758)

DEMETRIAS Bonelli, 1810
    Subgenus **RISOPHILUS** Leach, 1815
        AETOPHORUS Schmidt-Göbel, 1846
        imperialis (Germar, 1824)
    Subgenus **DEMETRIAS** Bonelli, 1810
        atricapillus (Linnaeus, 1758)
        monostigma Samouelle, 1819
            unipunctatus (Germar, 1824)

CYMINDIS Latreille, 1806
    Subgenus **CYMINDIS** Latreille, 1806
        axillaris (Fabricius, 1794)
        macularis Mannerheim in Fischer von Waldheim, 1824
    Subgenus **TARULUS** Bedel, 1906
        vaporariorum (Linnaeus, 1758)
            basalis Gyllenhal, 1810

PARADROMIUS Fowler, 1887
    linearis (Olivier, 1795)
    longiceps Dejean, 1826

DROMIUS Bonelli, 1810
    agilis (Fabricius, 1787)
        bimaculatus Dejean, 1825
    angustus Brullé, 1834
    meridionalis Dejean, 1825
        discus Puel, 1919
    quadrimaculatus (Linnaeus, 1758)

CALODROMIUS Reitter, 1905
    spilotus (Illiger, 1798)
        quadrinotatus (Zenker in Panzer, 1800) non (Fabricius, 1798)

PHILORHIZUS Hope, 1838
    DROMIOLUS Reitter, 1905
      melanocephalus (Dejean, 1825)
      notatus (Stephens, 1827)
        nigriventris (Thomson, C.G., 1857)
      quadrisignatus (Dejean, 1825)
      sigma (Rossi, 1790)
      vectensis (Rye, 1873)
        insignis sensu auctt. non (Lucas, 1846)

MICROLESTES Schmidt-Göbel, 1846
    BLECHRUS Motschulsky, 1847
      maurus (Sturm, 1827)
        glabratus sensu auctt. non (Duftschmid, 1812)
      minutulus (Goeze, 1777)
        glabratus (Duftschmid, 1812)

LIONYCHUS Wissmann, 1846
    quadrillum (Duftschmid, 1812)
      bipunctatus (Heer, 1838)
      unicolor Schilsky, 1888

SYNTOMUS Hope, 1838
    METABLETUS Schmidt-Göbel, 1846
      foveatus (Geoffroy in Fourcroy, 1785)
        foveola (Gyllenhal, 1810)
      obscuroguttatus (Duftschmid, 1812)
        atratus (Dejean, 1825)
      truncatellus (Linnaeus, 1761)

Tribe ODACANTHINI Laporte, 1834

ODACANTHA Paykull, 1798
    COLLIURIS sensu auctt. non De Geer, 1774
      melanura (Linnaeus, 1767)

Tribe DRYPTINI Schaum, 1857

DRYPTA Latreille, 1796
    dentata (Rossi, 1790)
      emarginata (Olivier, 1790)

Tribe ZUPHIINI Bedel, 1895

POLISTICHUS Bonelli, 1810
    POLYSTICHUS auctt. (misspelling)
      connexus (Geoffroy in Fourcroy, 1785)
        fasciolatus sensu auctt. non (Rossi, 1790)
      vittatus Brullé, 1834

Key to subfamilies of Carabidae

1. Scutellum hidden under hind margin of pronotum (Fig. 3); body form short and oval, with characteristic colour pattern (Plate 4) .... Omophroninae (p. 33)

- Scutellum visible; body shape and colouring different 2

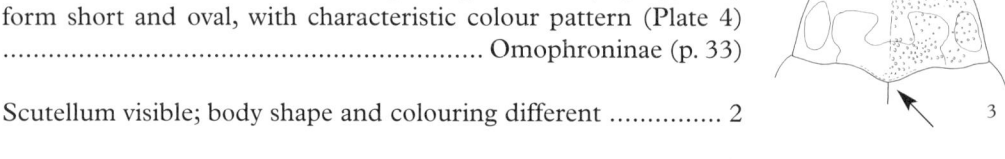

2. Antennae situated on dorsal surface of head behind the strongly toothed mandibles (Plates 1, 2; Fig. 4) .... Cicindelinae (p. 30)

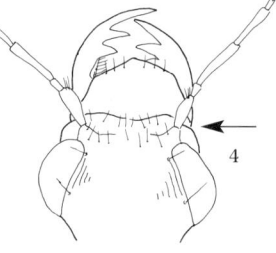

- Antennae situated laterally, outside the mandibles which have shorter or no teeth (e.g. Fig. 5) 3

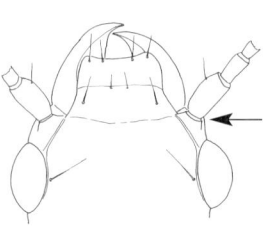

3. Abdomen with seven (female) or eight (male) ventral segments, including first divided segment (Fig. 6); elytra green or blue, apices truncate, pronotum and head red or orange (Plate 3) .... Brachininae (p. 32)

- Abdomen with only six ventral segments, including first divided segment (Fig. 2); colour and elytral apices usually not as above .... Carabinae (p. 33)

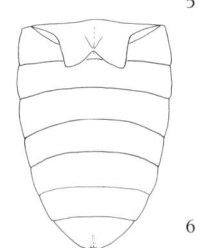

Subfamily CICINDELINAE Tiger beetles

A distinctive group of medium sized to large predators, many diurnally active. They were until recently considered to be a separate family from the Carabidae. Characterised by the antennae situated dorsally inside the outer margins of the mandibles, which are large, toothed internally and crossed in repose (Fig. 4). All the British species are diurnally active predators. Their predatory larvae lie in wait for their prey in vertical tunnels in the soil.

Key to genera of Cicindelinae

1. Pronotum clearly transverse (Plate 1), sides (proepisterna) with long white hairs .... 1. *Cicindela* Linnaeus (p. 31)

- Pronotum not wider than long (Plate 2), sides with few hairs .... 2. *Cylindera* Westwood (p. 32)

1. *CICINDELA* Linnaeus

There are about 40 European species in this major world-wide genus. Of the four British species, only *C. campestris* occurs in Ireland. All are winged and fly readily.

Key to species of *Cicindela*

1. Labrum black, with a sharp median keel (Fig. 7) ....
   .... 4. *sylvatica* Linnaeus

- Labrum at least partly yellow or brown, without keel 2

2. Elytra usually green, with discrete small pale spots (Plate 1) ....
   .... 1. *campestris* Linnaeus

- Elytra brown or coppery, with a transverse pale band medially (Figs 8, 10) 3

3. Transverse elytral band less narrowed and hardly angled medially (Fig. 8); apex of aedeagus curved in side view (Fig. 9) ....
   .... 2. *hybrida* Linnaeus

- Transverse elytral band narrowed and sharply angled medially (Fig. 10); apex of aedeagus straight in side view (Fig. 11) ....
   .... 3. *maritima* Latreille & Dejean

1. *Cicindela campestris* Linnaeus The Green Tiger Beetle Plate 1

Length 12-17 mm. Dorsal surface green, with coppery highlights; labrum and five or six spots on each elytron pale yellow. Appendages metallic coppery, bronze or purple. Head not wider than pronotum, which is transverse; labrum without central ridge. Elytral surface smooth with minute granules and punctures. Much of underside and base of legs with dense white hairs. There is a rare black form, var. *funebris* Sturm. In open grassland and moorland on sandy or peaty soils; iii-vi. Widespread and often abundant in Britain; more local and scarcer in Ireland.

2. *Cicindela hybrida* Linnaeus

Length 12-16 mm. Purplish-brown, with coppery and green metallic highlights visible under the microscope. Appendages black or dark brown with metallic reflections. Head as wide as pronotum which is transverse; labrum pale, without central ridge. Sides of thorax and underside of abdomen with dense white hairs. Elytra with three curved marks on each, the central band bent medially at about 45° (Fig. 8); surface with minute punctures. Aedeagus curved apically (Fig. 9), internal sac with two sub-apical teeth. On dunes, lowland heath and open grassland; v-viii. Very local, known only from Lancashire and south Cumbria; can be abundant.
Similar species: *C. maritima* (3) but in that species the central (angled) part of the transverse elytral band is narrower and almost parallel to the suture, whereas in *C. hybrida* this portion of the band is wider and inclined at about 45° to the suture. Other differences are in the aedeagi.

3. *Cicindela maritima* **Latreille & Dejean** The Dune Tiger Beetle

Length 11-15 mm. Central elytral band markedly narrowed and angled medially, inclined almost parallel to the suture (Fig. 10). Aedeagus straight apically (Fig. 11), without subapical teeth in internal sac. On coastal sand dunes; v-viii. Very local in Wales, north Devon, Kent and Norfolk; occasionally abundant.
Similar species: See *C. hybrida* (2).

4. *Cicindela sylvatica* **Linnaeus** The Wood Tiger Beetle

Length 14-18 mm. Dark brown, with outer edge of mandibles and elytral markings pale yellow. Elytral surface roughened, with irregular depressions as well as smaller punctures. Labrum black, with sharp central ridge (Fig. 7). Appendages black with metallic reflections. Elytral markings of basal and apical pale spots and central narrow curved band. Underside with dense white hairs. On lowland heaths; iv-vii. Very local in Dorset, Hampshire and Surrey; scarce.

2. CYLINDERA Westwood

This genus, with a single European species, was previously considered as a subgenus of *Cicindela*. Separated by the elongate and narrow body, with the pronotum not wider than long. Does not fly readily, unlike members of *Cicindela*. It does not occur in Ireland.

1. *Cylindera germanica* **Linnaeus** Plate 2

Length 9-12 mm. Body cylindrical, head wider than the quadrate pronotum, elytra narrowing basally, with shoulders hardly marked. Colour dark green; labrum, base of mandibles and three spots on each elytron pale yellow. Tibiae brown, antennae and remaining parts of legs black with metallic reflections. Underside glabrous except meso- and metasterna. On coastal grassy slopes and sand or silt near fresh water flushes; vi-ix. Very local, Dorset, Hampshire, Isle of Wight; occasionally abundant.

Subfamily BRACHININAE

The mainly tropical subfamily of Bombardier beetles are characterised by their extra abdominal segments (Fig. 6), and specialised defensive mechanism which can emit an audible release of spray of extremely irritant and corrosive vapour. The very large and widely distributed genus *Brachinus* Weber has about 40 European species. Only one is a permanent resident in Britain, although a second, *B. sclopeta* (Fabricius) occurs as an occasional immigrant. Neither occur in Scotland or Ireland.

Key to species of *Brachinus*

1. Elytra unicoloured, dark metallic blue or green ....
   .... 1. *crepitans* (Linnaeus)

- Elytra with a red-brown mark along basal half of intervals 1-2 ....
   .... 2. *sclopeta* (Fabricius)

Subgenus *Brachinus* Weber

1. *Brachinus crepitans* (Linnaeus) The Bombardier Beetle Plate 3

Length 6-9.5 mm. Head, pronotum and scutellum reddish brown, elytra dark metallic blue or green with fine pubescence. Appendages brown, except antennae darkened from third segment to apex. Head and pronotum narrow with scattered punctures, pronotum quadrate with sinuate sides. Elytra broad compared to forebody, widest behind middle, striae faint and unpunctured; apices transversely truncate, with a fringe of fine downwardly-directed setae. Abdomen with seven (female, Fig. 6) or eight (male) segments. Wings present. On chalky soils in grasslands, quarries and waste ground, where larvae are parasitic on pupae of other beetles, including Carabidae, Staphylinidae and Hydrophilidae; v-viii. Very local and often coastal in southern England and south Wales; can be abundant.
Similar species: Larger than *B. sclopeta* (2) with unicoloured elytra.

Subgenus *Brachynidius* Reitter

2. *Brachinus sclopeta* (Fabricius)

Length 4.5-7.5 mm. Head and pronotum pale reddish brown, elytra metallic blue with basal third or more of intervals 1-2 (including scutellum) reddish brown. Appendages uniformly pale yellow-brown. Head and pronotum narrow, pronotum slightly elongate, sides sinuate. Elytra broad and short, widest at or just behind middle, finely pubescent; apices without fringe of setae. Wings present. On waste ground, larvae possibly parasitic on other Carabidae. Known from very occasional specimens from Kent and Sussex until 1928; several found in London since 2005.
Similar species: See *B. crepitans* (1).

Subfamily OMOPHRONINAE

The single worldwide genus *Omophron* Latreille has one British species, immediately recognisable by its shape and colour pattern. It does not occur in Ireland.

1. *Omophron limbatum* Fabricius Plate 4

Length 5-6.5 mm. Body almost circular, pale yellow/brown with metallic green markings on pronotum and elytra. Appendages pale cream. Pronotum immovable, covering scutellum (Fig. 3); prosternum enlarged and covering mesosternum. Elytra with 15 striae. Wings present. At the sandy margins of standing fresh water; v-viii. Extremely local, in flooded gravel pits on the Kent/Sussex border, Norfolk and Suffolk; sometimes abundant.

Subfamily CARABINAE

This subfamily, as recognised here, comprises the vast majority of the Carabidae. They have antennae situated laterally outside the bases of the mandibles, a visible scutellum and only six ventral abdominal segments.

Key to tribes of Carabinae

1. Antennae with long setae, at least twice as long as the width of segments 2-6 (Fig. 12, Plate 29) .... Loricerini (p. 56)

- Antennae with short pubescence, or glabrous with shorter setae 2

2. Front tibia with two large apical spurs, its inner margin simple (Fig. 13) 3

- Front tibiae with a single large apical spur (rarely a small second spur) and either a sub-apical deep notch (Fig. 14) or an apical flat incision (Fig. 15) 6

3. Elytra without basal border (Fig. 16); length more than 10 mm 4

- Elytra with basal border (Fig. 17); length usually less than 10 mm 5

4. Head and mandibles very elongate with labrum deeply bilobed (Fig. 18, Plate 16); length less than 20 mm .... Cychrini (p. 44)

- Head and mandibles less elongate (except in one species longer than 25 mm), labrum not so deeply bilobed; length often more than 20 mm .... Carabini (p. 38)

5. Second elytral interval wider than other intervals; eyes very large (Plate 25) .... Notiophilini (p. 50)

- Second elytral interval of similar width to other intervals; eyes normal .... Nebriini (p. 45)

6. Front tibiae with an apical flat incision, without sub-apical notch (Fig. 19); eyes very large and convex, together wider than the frons (Plates 27, 28) .... Elaphrini (p. 54)

- Front tibiae with a sub-apical notch (Fig. 20); eyes normal or reduced 7

7. Scutellum in front of elytra on a waist or peduncle between elytra and pronotum (Fig. 21) 8

- Scutellum at least partly in line with anterior margins of elytra, forming a wedge between elytral bases (as in Figs 22,23) 9

8. Front legs fossorial, with long fixed apical spur and strong internal teeth (Fig. 24); length less than 6.5 mm .... Scaritini (p. 57)

- Front legs normal; length at least 6.5 mm .... Broscini (p. 62)

9. Apical segment of maxillary palpi very small, much shorter and narrower than penultimate segment (Fig. 25) ... Bembidiini (p. 69)

- Apical segment of maxillary palpi not shorter than, and usually not or hardly narrower than, penultimate segment 10

10. Head with deep semi-circular frontal furrows (Fig. 26); sutural stria usually recurved apically (Fig. 27) .... Trechini (p. 63)

- Head without such deep and curved frontal furrows; sutural stria usually simple 11

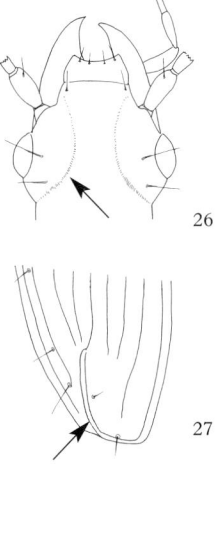

11. Outer edge of mandible with setiferous puncture (Fig. 28); length less than 10 mm 12

- Outer edge of mandible without setiferous puncture; length variable, many species more than 10 mm 13

12. Base of elytra with well-defined transverse border in front of the striae (Fig. 29) .... Pogonini (p. 102)

- Base of elytra without well-defined transverse border .... Patrobini (p.103)

13. Apex of mandibles widely truncate (Fig. 30), often notched above .... Licinini (p. 180)

- Apex of mandibles rounded or sharp, not notched above 14

14. Apex of elytra gradually rounded (as in Fig. 1), sometimes sinuate sub-apically (Fig. 31); apex of abdomen hardly extending beyond elytra except sometimes in gravid females (Plates 5-130) 15

- Apex of elytra transversely (Fig. 32) or obliquely (Fig. 33) truncate, apical abdominal tergite(s) (pygidium) often extending beyond elytra (Plates 131-147) 24

15. Head with two setiferous punctures just inside and behind each eye (Fig. 34) 16

- Head with only a single setiferous puncture inside each eye (Fig. 35) 20

16. Dorsal surface densely punctured and pubescent, eyes very prominent and elytra with characteristic red and black pattern (Plate 129) .... Panagaeini (p. 184)

- Without this combination of features 17

17. Eighth elytral stria strongly deepened, other striae very faint, elytral margins finely pubescent and length less than 2.5 mm (Plate 130) .... Perigonini (p. 185)

- Without this combination of features 18

18. Penultimate segment of labial palpi with three or more setae (Fig. 36) .... Zabrini (part – *Amara* and *Curtonotus*, p. 133)

- Penultimate segment of labial palpi with two or fewer setae 19

19. Apex of front tibiae strongly widened (Fig. 37); elytral margin usually crossing epipleuron near apex (Fig. 38) .... Pterostichini (p. 105)

- Apex of front tibiae narrower, more parallel-sided (Fig. 39); elytral margin simple apically (Fig. 40) .... Sphodrini + Platynini (p. 116)

20. Third antennal segment densely pubescent (except sometimes basally) 21

- Third antennal segment glabrous or, if pubescent, clearly less so than fourth segment 22

21. Upper surface densely pubescent and elytra yellow with 3 black spots (Plate 123) .... Chlaeniini (part – *Callistus*, p. 177)

- Without this combination of features .... Harpalini (p. 148)

22. Eighth elytral stria forming a keel apically (Fig. 41) .... Oodini (p. 179)

- Eighth elytral stria simple apically 23

23. Dorsal surface punctured and pubescent .... Chlaeniini (part – *Chlaenius*, p. 178)

- Dorsal surface unpunctured and glabrous .... Zabrini (part – *Zabrus*, p. 133)

24. Pronotum flattened dorso-ventrally, side margins distinct 25

- Pronotum cylindrical, without distinct side margins 27

25. Front tibiae with stout spines externally (Fig. 42); pronotum as wide as both elytra together (Plate 131) .... Masoreini (p. 185)

- Front tibiae with fine bristles or hairs; pronotum narrower than both elytra together 26

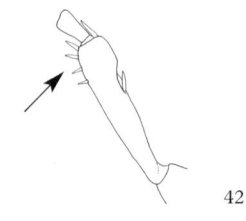

26. First antennal segment shorter than next two segments combined; elytral apices without membranous border .... Lebiini (p. 186)

- First antennal segment longer than next two segments combined; elytral apices with a fine membranous border (Plate 147, Fig. 43) .... Zuphiini (p. 199)

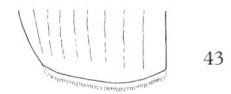

27. First antennal segment much longer than next three segments combined (Plate 146, Fig. 44) .... Dryptini (p. 199)

- First antennal segment shorter than next three segments combined (Plate 145) .... Odacanthini (p. 199)

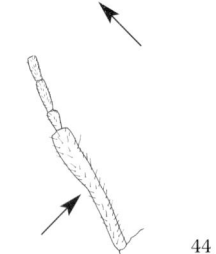

Tribe CARABINI

A worldwide tribe of large, predatory beetles. This and the following three tribes are characterised by the simple front tibiae with two apical spurs (Fig. 13). The Carabini are large active beetles, distinguished from Cychrini by their less specialised head and forebody and (usually) larger size.

Key to genera of Carabini

1. Hind angles of pronotum not produced backwards (Plate 5, Fig. 45) .... 1. *Calosoma* Weber (p. 38)

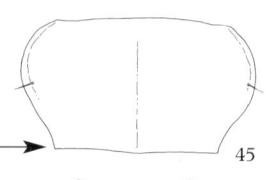

- Hind angles of pronotum produced backwards (Plates 6-15, Fig 46) .... 2. *Carabus* Linnaeus (p. 39)

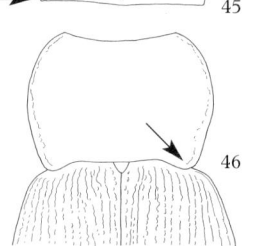

1. *CALOSOMA* Weber

A worldwide genus of about 120 species, six being European; only one is resident in Britain but a second is an occasional migrant. Separated from *Carabus* by the hind angles of the pronotum not extending backwards (Fig. 45) and by the presence of transverse furrows on the upper surface of the mandibles. Both species are winged and are known to fly. They are not found in Ireland.

Key to species of *Calosoma*

1. Length less than 23 mm; raised side border of pronotum absent behind middle (Plate 5, Fig. 47) .... 1. *inquisitor* (Linnaeus)

- Length more than 23 mm; pronotum with raised side borders throughout .... 2. *sycophanta* (Linnaeus)

1. *Calosoma inquisitor* (Linnaeus) Plate 5

Length 16-22 mm. Dark metallic bronze with green or purple reflections especially beneath and at the edges of elytra and pronotum. Appendages black. Pronotum transverse, with raised side borders only in front half (Fig. 47). Elytra with 15 striae and strong cross-striation, giving a rectangular pattern; fourth, eighth and 12th intervals with 8-10 strong setiferous punctures. In old oak woodland, where adults are arboreal, feeding on lepidopteran larvae; v –vi. Very local but widely distributed from southern England through mid Wales and Cumbria to Scotland. Usually scarce but occasionally abundant following outbreaks of caterpillar prey such as *Tortrix* species.

2. *Calosoma sycophanta* (Linnaeus)

Length 24-30 mm. Head and pronotum bluish metallic, elytra bright golden-green with coppery reflection. Sides of pronotum with raised borders throughout. Elytral punctures not conspicuous. An occasional immigrant into Britain; biology similar to *C. inquisitor* but in both deciduous and coniferous woods; v-viii. Southern half of England; occasional occurrences of single individuals.

2. *CARABUS* Linnaeus

An enormous group of about 700 species occurring throughout the Holarctic into the northern Oriental region. There are some 135 European species in many subgenera which are given generic status by some authors. Characterised by their large size and backwardly-protruding pronotal hind angles. The outer margin of the mid tibia is densely hairy in many species. They often show considerable intra-specific variation and there are numerous subspecific and varietal names. The 11 British (eight Irish) species are all in separate subgenera (as currently recognised) but these have not been used in the following key. For convenience the species are treated alphabetically rather than in taxonomic order of subgenera. Two species, *C. cancellatus* Illiger and *C. convexus* Fabricius have been found in the past; they are not resident and are not included in the key. They are wingless unless stated otherwise.

Key to species of *Carabus*

1. Elytra each with three raised keels but without obvious pits or granules (Plate 13) 2

- Elytra smooth (Plate 8) or irregularly sculptured (Plates 12, 14, 15) or with rows of granules or punctures as well as keels or low ridges (e.g. Plates 6, 7, 9, 11) 3

2. Length less than 20 mm; keels black .... 9. *nitens* Linnaeus

- Length 20 mm or more; keels metallic coloured ....
  .... 2. *auratus* Linnaeus

3. Head and pronotum very narrow, head elongate, pronotum quadrate (Plate 10); length usually more than 30 mm ....
   .... 6. *intricatus* Linnaeus

- Head quadrate, pronotum transverse; length less than 30 mm ... 4

4. Inner surface of penultimate segment of labial palpi with two (rarely three on one side) setae ('bisetose', Fig. 48) 5

- Inner surface of penultimate segment of labial palpi with three to five setae ('multisetose', Fig. 49) 10

5. Elytra each with three strong raised keels, with single intervening rows of either pits or elongate granules (Plates 7, 9) 6

- Elytra either smooth or with finer and more numerous ridges, granules or striae 7

6. Intervals between keels each with 8-10 deep pits (Plate 7); length more than 23 mm .... 3. *clatratus* Linnaeus

- Intervals between keels each with 8-12 elongate raised granules (Plate 9); length less than 23 mm .... 5. *granulatus* Linnaeus

7. Elytra without obvious sculpturing or punctures; body all black and very convex (Plate 8) .... 4. *glabratus* Paykull

- Elytra with evident sculpturing or rows of punctures; usually at least partly metallic, less convex 8

8. Elytra with indistinct granular sculpturing and three rows of fine punctures (Plate 12) .... 8. *nemoralis* Müller

- Elytra with distinct rows of fine ridges and granules (Plates 6, 11)
  9

9. Length less than 21 mm; with two or three irregular elytral ridges between each row of granules (Plate 6) .... 1. *arvensis* Herbst

- Length more than 21 mm; with three regular and equal-sized ridges between each row of granules (Plate 11) .... 7. *monilis* Fabricius

10. Elytra with fine but evident elongate, irregular sculpturing (Plate 14); elytra less elongate, length/maximum breadth less than 1.6; hind angles of pronotum strongly raised above rather flat disc (side view Fig. 50) .... 10. *problematicus* Herbst

- Elytra almost smooth, or occasionally with fine longitudinal striations (Plate 15); elytra more elongate, length/maximum breadth more than 1.7; hind angles of pronotum not raised above the level of the convex disc (side view Fig. 51) .... 11. *violaceus* Linnaeus

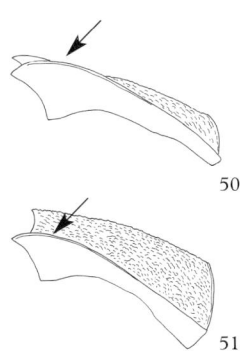

Subgenus *Eucarabus* Géhin

1. *Carabus arvensis* Herbst Plate 6

Length 16-20 mm. Colour variable, typically coppery but can be green, purple, bluish or even black; males noticeably more brightly coloured than females. Appendages black, labial palpi with two setae on penultimate segment. Pronotum rather flat, sides rounded and contracted almost evenly, hind angles level or slightly down-turned. Elytra with rows of major granules separated by two or three irregular low ridges, these often interrupted, or transversely sculptured. The British and Irish form is subspecies *sylvaticus* Dejean. On heaths and moorland; v-vii. Widespread in most of Britain and Ireland except in central and eastern England; often abundant.

Similar species: *C. granulatus* (5) but *C. arvensis* has a flatter pronotum with hind angles down-turned, and with at least double rows of low granules or striae between the major rows, which are less pronounced than in *C. granulatus*. *C. monilis* (7) has similar coppery colour and rows of elytral granules but is larger and more elongate, with more regular elytral sculpturing of three regular and equal-sized ridges between each row of granules. The elytral ridges are more complete than in *C. problematicus* (10).

Subgenus *Autocarabus* Seidlitz

2. *Carabus auratus* Linnaeus

Length 20-27 mm. Metallic green or golden, sides of elytra coppery. Legs, palpi and first four antennal segments brown. Pronotum transverse, rather flat, sides evenly rounded. Elytra with three uninterrupted smooth keels that are concolorous with the rest of the dorsal surface. In gardens and suburban habitats; v-viii. Introduced and established in Berkshire and north Essex; scarce.

Similar species: *C. nitens* (9), the only other species with smooth, keeled elytra is smaller, with keels and all appendages black.

Subgenus *Limnocarabus* Géhin

3. *Carabus clatratus* Linnaeus Plate 7

Length 22-30 mm. Black usually with a green/golden reflection, elytral foveae coppery. Appendages black, penultimate segment of labial palpi with two setae. Pronotum transverse, sides sinuate basally, rather flat except for two deep depressions inside and in front of hind angles. Elytra with three uninterrupted keels, the innermost lying along suture for most of its length; intervals between keels each with 8-10 large foveae which become indistinct at base, apex and sides. Wings present or absent, sometimes capable of flight. In bogs, on wet moorland and at margins of fresh water; v-vii. In north west Scotland and throughout Ireland. The Scottish form is generally smaller and less brightly coloured, with the rows of elytral foveae less distinct and even becoming granular apically; usually scarce.

Subgenus *Oreocarabus* Géhin

4. *Carabus glabratus* Paykull Plate 8

Length 22-30 mm. Entirely black, with at most a slight metallic blue reflection. Labial palpi bisetose. Pronotum broad with a wide side border; hind angles raised at least to level of disc. Elytra shining with a fine, even microsculpture of small punctures connected by a network of fine lines. The British and Irish form is subspecies *lapponicus* Born. On moorland and in forests; v-viii. Northern England, north Wales, Scotland and Ireland; can be abundant.
Similar species: Distinguished from black examples of *C. violaceus* (11), the only other species with smooth elytra, by its wider and much more convex body, as well as by the bisetose labial palpi.

Subgenus *Carabus sensu stricto*

5. *Carabus granulatus* Linnaeus Plate 9

Length 16-23 mm. Black, usually with a coppery or greenish reflection. Appendages black, labial palpi bisetose. Pronotum transverse, sides usually contracted basally and clearly sinuate; basal foveae deep; side borders and hind angles raised. Elytra with three keels separated by single rows of 8-12 elongate granules; surface with rough microsculpture between these. Wings present or absent but flight unknown in Britain. In open very wet habitats such as flood plains, marshes and bogs in Britain. In Ireland found in almost all open habitats; iv-vi. The British form is subspecies *granulatus* Linnaeus, whereas in Ireland the species is subspecies *hibernicus* Lindroth. This has shallower elytral keels and stronger microsculpture. Widespread throughout Britain, except in the north-east of Scotland, and Ireland; often abundant.
Similar species: See *C. arvensis* (1). The elytral ridges and granules are more complete than in *C. problematicus* (10).

Subgenus *Chaetocarabus* Thomson C.G.

6. *Carabus intricatus* Linnaeus The Blue Ground Beetle Plate 10

Length 25-35 mm. Bright metallic blue, with irregular elongate elytral sculpturing. Appendages longer than in all other British *Carabus*. Penultimate segment of labial palpi multisetose, apical segments of all palpi exceptionally expanded and hatchet-shaped. Head and pronotum elongate, pronotum almost flat. Elytra large and rather flat, apical margins sinuate, especially in female. In mossy deciduous pasture woodlands in river valleys; i-xii. Extremely local in Devon and Cornwall only; very scarce.

Similar species: Larger and flatter than *C. problematicus* (10), the only other species that has similar elongate elytral sculpture and (sometimes) blue colour.

Subgenus *Morphocarabus* Géhin

7. *Carabus monilis* Fabricius Plate 11

Length 22-26 mm. Coppery or greenish metallic. Pronotum wide, sides contracted basally with side borders and hind angles clearly raised almost to level of disc. Elytra elongate, with rows of low granules, separated by three regular and equal-sized ridges. In cultivated fields, grasslands and scrub; v-vii. Central and southern England, and eastern Wales; once quite abundant but now usually scarce; there are only pre-1900 records from Ireland.

Similar species: Elytra as long as in *C. violaceus* (11) but sculpturing more regular. See also *C. arvensis* (1).

Subgenus *Archicarabus* Seidlitz

8. *Carabus nemoralis* Müller Plate 12

Length 20-26 mm. Head and pronotum black, margins of pronotum with a purple or coppery reflection; elytra black usually with a greenish metallic reflection. Appendages black, penultimate segment of labial palpi with two setae. Pronotum wide, sides contracted basally and slightly sinuate, hind angles down-turned. Elytra rather convex, with rough, transverse scale-like sculpturing, interrupted by three rows of 8-10 small but distinct punctures. In gardens, fields and most habitats that are not exceptionally wet; iv-vi, ix-x. Widespread in most of Britain and Ireland; abundant.

Similar species: With less elytral sculpturing than *C. problematicus* (10), more convex and wider than *C. violaceus* (11) and separated from both by the bisetose labial palpi and rows of punctures on the elytra.

Subgenus *Hemicarabus* Géhin

9. *Carabus nitens* Linnaeus Plate 13

Length 13-18 mm. Head black with a coppery or green reflection; pronotum metallic coppery red sometimes with a green reflection; elytra metallic green with coppery red margins and black keels. Occasionally all black. Appendages black, penultimate segment of labial palpi with two setae. Front tibiae with an apical projection. Pronotum domed centrally with sides rounded and side borders hardly raised. Elytra with three more or less complete keels, the outer two sometimes interrupted by small breaks or punctures but not forming regular lines of granules; surface between keels with transverse sculpturing. In wet lowland heaths, dune slacks and upland *Sphagnum* bogs, usually where there is some bare ground; v-vi. Very local in southern England (Hampshire and Dorset), local in northern England, the southern half of Scotland and the northern half of Ireland; usually scarce.

Similar species: See *C. auratus* (2).

Subgenus *Mesocarabus* Thomson C.G.

10. *Carabus problematicus* Linnaeus Plate 14

Length 20-28 mm. Black, usually with blue or violet metallic reflection, over most of dorsal surface. Appendages black, penultimate segment of labial palpi multisetose. Pronotum barely transverse, sides contracted and sometimes sinuate basally with hind angles distinctly raised above the level of rather flat disc (Fig. 50). Elytra shining, with distinct sculpture of longitudinal granulation, the larger granules sometimes forming interrupted keels, separated by three rows of smaller granules. The British and Irish form is subspecies *harcyniae* Sturm. In woodland, rough grasslands and moorlands; v-viii. Widespread in Britain, especially in the north and west, and in Ireland on the hills; often abundant.
Similar species: Elytra less elongate and flatter than in *C. violaceus* (11), with rougher elytral sculpturing; elytra more extensively blue; pronotum less transverse, and flatter. See also *C. arvensis* (1), *C. granulatus* (5), *C. intricatus* (6) and *C. nemoralis* (8).

Subgenus *Megodontus* Solier

11. *Carabus violaceus* Linnaeus The Violet Ground Beetle Plate 15

Length 20-30 mm. Dull black with margins of pronotum and elytra metallic, usually violet, sometimes blue or greenish. Appendages black; penultimate segment of labial palpi multisetose. Pronotum clearly transverse, with hind angles hardly raised, lower than level of the disc (Fig. 51). Elytra with fine granulation, sometimes arranged into longitudinal rows but without or with only a suggestion of raised lines. The usual British form is subspecies *sollicitans* Hartert; subspecies *purpurascens* Fabricius has more elongate elytra with distinct longitudinal striations. In woodland, gardens, moorlands and most undisturbed habitats; iv-ix but usually viii-ix as it is an autumn breeder. Widespread in Britain except at high altitudes and in the extreme north; often abundant. *C. violaceus purpurascens* only occurs in the extreme south of England. Neither subspecies occurs in Ireland.
Similar species: See *C. glabratus* (4), *C. monilis* (7), *C. nemoralis* (8) and *C. problematicus* (10).

Tribe CYCHRINI

A small tribe of four genera of specialised snail-hunters, adapted for inserting their foreparts into the shells of their prey. There is a single European genus, *Cychrus* Fabricius, with 15 European species of which only one occurs in Britain and Ireland.

1. *Cychrus caraboides* (Linnaeus) Plate 16

Length 14-19 mm. Dull black with fine granulate sculpturing. Appendages black. Head narrow and elongate, with long protruding mandibles and deeply bilobed labrum (Fig. 18). Pronotum narrow, quadrate with flat dorsal surface; hind angles rounded, not produced backwards. Elytra ovoid, very convex dorsally, sometimes with a hint of three longitudinal rows of granules. Wings absent. Stridulates loudly if handled. In woods and upland grasslands and (especially in Ireland) on peaty moors; vi-ix. Throughout Britain and Ireland; abundant.

Tribe NEBRIINI

A Holarctic tribe of about 650 species, with over 100 in Europe. Mostly rather flat species with strongly transverse, cordate pronotum and long, slender legs without subapical notch. Britain has 13 species in four genera.

Key to genera of Nebriini

1. Elytra with large pores on intervals 3 and 5 (Plate 24); scutellary stria extending almost to apex of elytra ... 4. *Pelophila* Dejean (p. 50)

- Elytra without large pores; scutellary stria very short (Fig. 52) ... 2

2. Outer sides of mandibles expanded (Fig. 53) .... 1. *Leistus* Frölich (p. 45)

- Outer sides of mandibles not expanded 3

3. Head dark brown or black, elytra unicolourous (Plates 21, 22) or black with red-brown sides and apices (Plate 20) .... 2. *Nebria* Latreille (p. 48)

- Head pale; elytra bicoloured, either black with pale cream margins, or pale with black markings (Plate 23) .... 3. *Eurynebria* Ganglbauer (p. 50)

1. *LEISTUS* Frölich

Flat, long-legged carabids, mostly smaller than *Nebria*; mouthparts adapted for catching Collembola, with expanded mandibles (Fig. 53), a ventral setal cage beneath the head, and a projecting, toothed labial ligula. There are 36 European species; the six British species are in three subgenera. Only three species occur in Ireland

Key to species of Leistus

1. Side borders of pronotum wider than width of second antennal segment (Fig. 54) 2

- Side borders of pronotum not wider than width of second antennal segment (Figs 55-57) 4

2. Pronotum nearly twice as wide as long (Fig. 54); body black or dark brown with paler side margins ...... 2. *rufomarginatus* (Duftschmid)

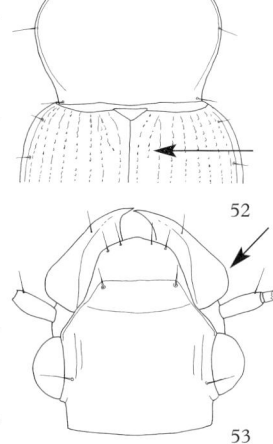

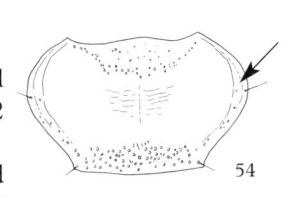

- Pronotum about 1.5 times as wide as long; body at least partly bluish-metallic (e.g. Plate 17), sides not paler 3

3. Pronotal epipleura brown, paler than pronotum; femora mostly reddish brown, sometimes darker apically ... 1. *montanus* Stephens

- Pronotal epipleura metallic bluish, as dark as dorsal surface of pronotum; femora dark brown or black ... 3. *spinibarbis* (Fabricius)

4. Dorsal surface dark brown with bluish metallic reflection (Plate 18); side borders of pronotum with small punctures (Fig. 55) .... 4. *fulvibarbis* Dejean

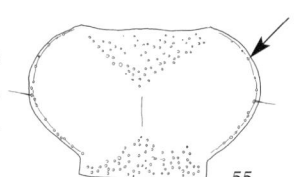

- Dorsal surface mostly reddish brown, non-metallic (Plate 19); side borders of pronotum unpunctured (Figs 56,57) 5

5. Head as pale as pronotum; sides of pronotum abruptly sinuate in front of right-angled hind angles (Fig. 56) .... 5. *ferrugineus* (Linnaeus)

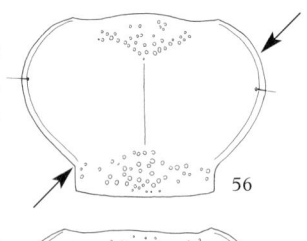

- Head darker than pronotum; sides of pronotum slightly sinuate in front of obtuse hind angles (Fig. 57) .... 6. *terminatus* (Hellwig in Panzer)

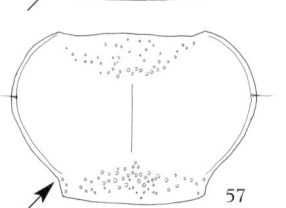

Subgenus *Pogonophorus* Latreille

1. *Leistus montanus* Stephens

Length 7-9.5 mm. Greenish or bluish metallic with sides of pronotum and elytra reddish; epipleura reddish brown. Antennae pale brown, third segment almost as long as fifth; legs very long, mostly pale brown, apex of femora sometimes darker. Sides of frons in front of eyes with both punctures and irregular wrinkles. Pronotum strongly sinuate in front of sharp, setose hind angles. Elytra twice as long as wide, sides gradually widened behind toothed shoulders. Wings absent. On well drained mountain soils and scree; vii-ix. North Wales, Cumbria, west of Scotland, Ireland mainly in the south; very scarce.
Similar species: *L. spinibarbis* (3) is on average slightly larger and darker coloured, purplish blue rather than greenish, with darker femora and epipleura.

2. *Leistus rufomarginatus* (Duftschmid)

Length 8-9.5 mm. Dark brown without metallic reflection, margins of pronotum and elytra and epipleura pale brown. Antennae pale, with third segment shorter than the fifth; legs pale brown. Sides of frons almost smooth. Pronotum very transverse, with wide border, slightly sinuate immediately in front of the setose hind angles (Fig. 54). Elytra twice as long as wide, sides abruptly widened behind the toothed shoulders. Wings present. In gardens, deciduous woods and conifer plantations; v-viii. An immigrant species that is now widespread throughout England, Wales and local in southern and central Scotland; initially abundant but now scarcer since it has become established.

Similar species: *L. spinibarbis* (3), the only other species of similar size, is metallic blue. *L. fulvibarbis* (4) is smaller, with a bluish reflection, and narrower pronotum with a narrow side border.

3. *Leistus spinibarbis* (Fabricius) Plate 17

Length 8-10.5 mm. Black with a strong bluish (occasionally greenish or violet) reflection. Antennae with four basal segments dark brown to black; legs mid brown with femora dark brown. Sides of frons wrinkled with few or no punctures. Pronotum 1.5 times as wide as long, hind angles sharp, setose. Elytra less than twice as long as wide, sides abruptly widened behind the toothed shoulders. Wings present. In woods, gardens and near the coast; vi-ix. Widespread in England and Wales, very local in Scotland; abundant except in the north of its range.
Similar species: See *L. montanus* (1) and *L. rufomarginatus* (2). Larger than *L. fulvibarbis* (4), with a brighter blue reflection and darker femora.

Subgenus *Leistophorus* Reitter

4. *Leistus fulvibarbis* Dejean Plate 18

Length 6.5-8.5 mm. Dark brown with a bluish metallic reflection. Appendages pale to mid brown. Sides of pronotum abruptly sinuate in front of right-angled hind angles which are without setae; side borders narrow and finely punctured (Fig. 55). Elytra without shoulder tooth. Wings present. In woodlands with damp litter; vi-xi. Widespread in England, Wales and Ireland, local in Scotland; abundant.
Similar species: See *L. spinibarbis* (3). Pale examples, lacking the metallic reflection, can be distinguished from *L. ferrugineus* (5) and *L. terminatus* (6) by the fine punctures along the side borders of the pronotum and the more prominent elytral shoulders

Subgenus *Leistus sensu stricto*

5. *Leistus ferrugineus* (Linnaeus) Plate 19

Length 6-8 mm. Pale to mid brown, head not darker than pronotum. Appendages light brown. Side border of pronotum very narrow, unpunctured except for the large setae at middle of each side, sharply sinuate in front of the right-angled hind angles (Fig. 56). Sides of elytra strongly widened behind the non-toothed shoulders. Wings absent. In fields, gardens and open, moderately dry places; vi-ix. Widespread in England, local in Wales and Scotland; abundant in the south and east, scarce elsewhere.
Similar species: See *L. fulvibarbis* (4). Distinguished from *L. terminatus* (6) by having a uniformly pale brown dorsal surface and by the right-angled hind pronotal angles.

6. *Leistus terminatus* (Hellwig in Panzer)

Length 6-8 mm. Pale to mid brown, with head, apical part of elytra and abdomen dark brown to almost black. Appendages pale brown. Sides of pronotum only slightly sinuate in front of obtuse hind angles (Fig. 57). Sides of elytra strongly widened behind the non-toothed shoulders. Wings present or absent. In damp grasslands, woodland and gardens; vi-ix. Widespread in both Britain and Ireland; very abundant.
Similar species: See *L. fulvibarbis* (4) and *L. ferrugineus* (5).

2. *NEBRIA* Latreille

A large genus of more than 80 European species, with five British species (three in Ireland) in three subgenera. They are larger than *Leistus*, with less specialised mouthparts, shorter appendages and less cordate pronotum.

Key to species of *Nebria*

1. Elytra uniformly black, dark brown or reddish; pronotum unicoloured 2

- Elytra black with pale margins; pronotum pale with front and hind margins darker (Plate 20) .... 1. *livida* (Linnaeus)

2. Shoulder angle of elytra sharp (Fig. 58); antennae brown; metepisterna and sides of first abdominal sternite punctured 3

- Shoulder angle of elytra rounded (Fig. 59); at least segments 3-4 of antennae black; metepisterna and abdominal sternites unpunctured 4

3. Dorsal surface of hind tarsi with fine pale hairs (Fig. 60) ....
   .... 2. *brevicollis* (Fabricius)

- Dorsal surface of hind tarsi glabrous ....
   .... 3. *salina* Fairmaire & Laboulbène

4. Second antennal segment with two apical setae (Fig. 61) ....
   .... 4. *nivalis* (Paykull)

- Second antennal segment with one or no setae ....
   .... 5. *rufescens* (Ström)

Subgenus *Paranebria* Jeannel

1. *Nebria livida* (Linnaeus) Plate 20

Length 13-16 mm. Head black; pronotum red-brown with anterior and posterior margins black; elytra black with red-brown margins and apices. Appendages pale brown. Penultimate segment of labial palpi with only two (apical) setae. Pronotum widest in front third, strongly contracted to obtuse hind angles, which hardly protrude. Third elytral interval with dorsal punctures. Wings present. At the base of clay or sandy cliffs or banks near water; viii-ix. Very local and mostly coastal in eastern England from Norfolk to Yorkshire; very scarce.

Subgenus *Nebria sensu stricto*

2. *Nebria brevicollis* (Fabricius) Plate 21

Length 11-14 mm. Black or dark brown. Appendages dark to mid brown, femora sometimes almost black; dorsal surface of all tarsi with fine pale hairs (Fig. 60). Pronotum cordiform, strongly punctured laterally and across the base. Elytra with distinct shoulder angle (Fig. 58) and with 5-7 dorsal punctures on the third interval; microsculpture extremely fine and transverse, just visible at x50. Metepisterna and margins of first abdominal segment coarsely punctured. Wings present. In almost all habitats, especially woodland, gardens and agricultural grasslands; iv-v, viii-x. Widespread throughout Britain and Ireland except at high altitudes; often extremely abundant.
Similar species: *N. salina* (3) has no hairs on the dorsal surface of the hind tarsi and coarser, less transverse elytral microsculpture.

3. *Nebria salina* Fairmaire & Laboulbène

Length 11-14 mm. Black or dark brown. Appendages mid to dark brown; dorsal surface of hind tarsi without hairs. Elytral microsculpture consisting of slightly transverse reticulation, visible at x30. Wings present. On sandy or unproductive soils and lowland heaths; v, viii-ix. Local throughout Britain and Ireland, except the north-west of Scotland where it is widespread; sometimes abundant.
Similar species: See *N. brevicollis* (2).

Subgenus *Boreonebria* Jeannel

4. *Nebria nivalis* (Paykull)

Length 9-11 mm. Black, rather elongate. Antennae black except extreme apical segments sometimes paler; apex of second segment with two setae (Fig. 61). Legs mid brown to almost black but with femora always darker apically than basally. Side borders of pronotum somewhat rugose but hardly punctured; minute ridge present inside lateral seta. Elytra widened posteriorly, with dorsal punctures on the third, and often also the fifth, intervals; shoulders rounded. Wings present. On barren and exposed hill tops, mostly at high altitudes; vi-viii. Very local in north and west Scotland, north-west England and north Wales; very scarce.
Similar species: Distinguished from *N. rufescens* (5) by having two apical setae on the second antennal segment, as well as a minute ridge inside the lateral pronotal seta and the femora darkened apically.

5. *Nebria rufescens* (Ström) Plate 22

Length 9-11 mm. Black (form *gyllenhali* Schönherr) or with elytra reddish (type form). Antennae black, sometimes with basal two segments and apices brown; legs usually all dark brown to black. Form *balbii* Bonelli has red-brown legs but dark antennae. Pronotum with side borders finely punctured, without ridge inside lateral seta. Elytra almost parallel sided, with dorsal punctures only on the third interval; shoulders rounded (Fig. 59). Wings present. On moorland, upland heaths and by stony river and stream margins; vi-viii. Widespread in northern and western Britain and Ireland, local in the extreme south-west of England; often abundant.
Similar species: See *N. nivalis* (4).

3. *EURYNEBRIA* Ganglbauer

A single distinctive species sometimes included within *Nebria*. It differs in having a multisetose penultimate segment of the labial palpi and lack of dorsal elytral punctures.

1. *Eurynebria complanata* (Linnaeus) Plate 23

Length 16-23 mm. Pale cream to golden brown with two transverse bands of elongate black markings on elytra. Antennae pale, very long, basal four segments glabrous, remainder with fine pubescence; legs pale brown or cream. Penultimate segment of labial palpi with numerous setae along inner surface. Pronotum cordiform, with a transverse depression in front of the bisinuate hind margin; posterior angles slightly acute and produced backwards. Wings present. On sandy beaches above the high tide line; v, viii-ix. Very local in south-west England and south-east Ireland; sometimes abundant.

4. *PELOPHILA* Dejean

A single European species, that also occurs in north America. It is more convex than *Nebria*, and distinguished by the long elytral scutellary stria and foveate punctures. Superficially similar to the elaphrine *Blethisa multipunctata*.

1. *Pelophila borealis* (Paykull) Plate 24

Length 9-12 mm. Rather convex, dark metallic bronze, occasionally bluish or greenish. Antennae black or with bases of some or all segments brown, less elongate than in *Nebria*; legs black or brown. Head unpunctured between eyes, with irregular elongate wrinkling inside each eye. Pronotum cordate, with punctuation along the side borders and in both the anterior and posterior transverse depressions; with a deep fovea inside each of the slightly acute hind angles. Elytra with all striae reaching almost to apex, without abbreviated scutellary stria; third and fifth intervals with about six and three large foveate punctures, respectively. Wings present. Near standing lowland water on open, muddy ground with both stones and vegetation; also by vegetated flushes on high ground; v-vi. Very local, Orkney and Shetland, with one known recent mainland Scottish site. Widespread in north and west Ireland; can be abundant.

Tribe NOTIOPHILINI

A very distinctive group of small, diurnal beetles, with large eyes and widened second elytral interval, adapted for active hunting of prey (mostly Collembola). The single genus *Notiophilus* has 14 European species, of which eight occur in Britain, six in Ireland.

Key to species of *Notiophilus*

1. Second elytral interval wider, at least as wide as intervals 3-5 combined (Plate 25, Fig. 62) 2

- Second elytral interval narrower, about as wide as intervals 3-4 combined (Fig. 63) 5

2. At least femora brown metallic or black; basal antennal segments darkened dorsally; elytra with yellow apical patch (Plate 25, Fig. 62) 3

- Legs and base of antennae entirely pale brown, non metallic; elytra without distinct yellow apical patch .... 7. *rufipes* Curtis

3. Striae fine and shallow, intervals 3 outwards dull and flat, with distinct microsculpture (Fig. 62) .... 8. *substriatus* Waterhouse

- Striae deeper, all intervals shiny and convex (Plate 25) 4

4. Sides of pronotum more sinuate, frons with six coarse ridges (Fig. 64); fourth elytral interval usually with only a single puncture (Plate 25) .... 3. *biguttatus* (Fabricius)

- Sides of pronotum almost straight, frons with 7-10 fine ridges (Fig. 65); fourth elytral interval usually with two punctures .... 6. *quadripunctatus* Dejean

5. Legs entirely black 6

- Tibiae reddish or brown, paler than femora 7

6. Elytral apices almost flat, with two apical punctures (side view Fig. 66); striae very finely punctured .... 1. *aesthuans* Motschulsky

- Elytral apices convex, usually with a single apical puncture (side view Fig. 67); striae more deeply punctured ... 2. *aquaticus* (Linnaeus)

7. Frontal furrows deep and parallel, back of frons with a wide unpunctured area (Fig. 68) .... 4. *germinyi* Fauvel

- Frontal furrows shallow and converging towards the rear, where there is only a narrow unpunctured area (Fig. 69) .... 5. *palustris* (Duftschmid)

1. *Notiophilus aesthuans* **Motschulsky**

Length 4.2-5.3 mm. Shining dark bronze, legs and antennae (except sometimes segments 2-3) black. Frons with six almost parallel ridges, hind margin almost unpunctured. Pronotum moderately narrowed posteriorly, all borders with fine punctures mostly separated by at least their own width. Elytra with second interval twice as wide as third, fourth with a single puncture; all striae extremely fine, absent from apical quarter of elytra; apex almost flat and finely reticulate, with two punctures inside two pre-apical ridges (Fig. 66). Wings present. On dry, open and well-drained habitats such as gravel workings, exposed riverine sediments and old mine workings; vi-x. Very local, mostly in northern England and Scotland; a single Irish record from Down; usually very scarce.
Similar species: Smaller, flatter, more finely punctured and shinier than *N. aquaticus* (2), which is the only other species with a narrow second elytral interval and all black legs. *N. aquaticus* usually has only a single apical elytral puncture.

2. *Notiophilus aquaticus* **(Linnaeus)**

Length 5-6 mm. Bronze to black, moderately shining, apex of elytra sometimes slightly paler. Antennae black with second, third and base of fourth segments dark brown; legs black. Frons with six ridges which are parallel or slightly converging posteriorly, hind margin slightly wrinkled laterally. Pronotum moderately narrowed behind, all borders with coarse punctures often closer together than their own width. Elytra rather narrow, more than 1.5 times as long as combined width, widest behind middle; striae moderately punctured, visible almost to apex, which is rounded in side view, with usually a single apical puncture inside the pre-apical ridges (Fig. 67); very rarely a second apical puncture is also present. Wings present or absent. In open habitats such as grassland, dunes, moorland and by rivers; v-ix. Widespread throughout Britain and Ireland; often abundant especially in the north.
Similar species: See *N. aesthuans* (1). Recognisable from all other species in the genus by the completely dark legs.

3. *Notiophilus biguttatus* **(Fabricius)** Plate 25

Length 5-6 mm. Bright bronze or coppery, elytra with distinct apical yellow spot extending one quarter or more the length of each elytron. Antennae dark with second to fourth segments pale; legs with femora and tarsi black or dark metallic, tibiae yellow to brown. Frons with five to six rather coarse ridges, these sometimes converging posteriorly. Pronotum strongly and coarsely punctured except centrally, with sides slightly contracted and sinuate posteriorly (Fig. 64). Elytra with second interval at least three times as wide as each of third and fourth intervals, which are hardly widened; striae coarse, as wide as third interval; fourth interval with (usually) a single prominent setiferous puncture; all intervals shining. Wings present or absent. In gardens, woodland, grasslands and arable land; iv-x. Ubiquitous throughout Britain and Ireland; often very abundant.
Similar species: *N. quadripunctatus* (6) has straighter sides to the pronotum, and at least seven finer frontal ridges, as well as (usually) two punctures on the third elytral interval. *N. substriatus* (8), the only other species with pale apical elytral spots, has much finer striae, with third interval outwards distinctly sculptured.

4. *Notiophilus germinyi* **Fauvel**

Length 4.5-5.5 mm. Bright coppery or bronze to almost black, sometimes with bluish or purple reflections, especially on the head. Antennae black with basal four segments paler

but first segment at least partially darkened above; legs dark except tibiae yellow-brown to dark red-brown. Head not wider than pronotum; frons with six straight, parallel ridges and a wide unpunctured zone behind these (Fig. 68). Pronotum rather arched, and strongly narrowed behind; strongly punctured except on disc. Elytra slightly widened behind; second interval as wide as next two combined (Fig. 63); outer intervals with very fine reticulation; apical pale spot absent. Wings present or absent. On moorland, heaths and dry grasslands; iv-viii. Widespread throughout Britain and Ireland; abundant.
Similar species: *N. palustris* (5) has frontal ridges that converge behind, with only a narrow unpunctured zone; also the outer elytral intervals are totally smooth in that species.

5. *Notiophilus palustris* (Duftschmid)

Length 4.5-5.5 mm. Dark coppery or bronze to black. Antennae black with segments two-four paler; legs dark with paler tibiae. Head slightly wider than pronotum; frons with five to six sometimes quite shallow ridges which converge behind; back of head with only a narrow unpunctured zone (Fig. 69). Pronotum strongly narrowed behind and elytra narrowed in front to meet rear of pronotum. Second elytral interval as wide as next two combined, outer intervals completely smooth and shiny. In damp grasslands, woodland on heavy soils and other shaded habitats; v-ix. Widespread in England, local in Scotland and Ireland; sometimes abundant.
Similar species: See *N. germinyi* (4).

6. *Notiophilus quadripunctatus* Dejean

Length 4.8-5.5 mm. Coppery or bronze with yellow apical spot on elytra. Legs dark with paler tibiae. Frons with seven to ten fine, sharp ridges which converge slightly behind. Pronotum rather flat, strongly punctured except on disc, narrowing posteriorly but with sides almost straight (Fig. 65). Elytra narrow, with coarse striae, second interval as wide as next three, fourth interval usually with two prominent setiferous punctures. Wings present or absent. On dry, sandy heaths, gravel pits and other open, well drained habitats; v-x. Very local in southern England, rarer in the north and Wales, only once recorded from Scotland; very scarce.
Similar species: See *N. biguttatus* (3).

7. *Notiophilus rufipes* Curtis

Length 5.5-6.5 mm. Bright coppery or bronze to almost black. Antennae with basal four segments pale; legs very long, pale yellowish brown, only femora with a darker metallic reflection. Frons with six to eight rather fine converging ridges. Pronotal disc shining, sides quite strongly sinuate behind. Elytra with second interval as wide as next three combined, striae coarsely punctured but narrower than third interval; apex with only a hint of yellowish colour. Wings present. In gardens and deciduous woodland where there is leaf litter; v-ix. Widespread in south-east England, local in the west and north; the only Scottish record is from Orkney; abundant in the south-east, scarce elsewhere.
Similar species: Can be distinguished from all other species of the genus by the pale, long legs and antennal base.

8. *Notiophilus substriatus* Waterhouse

Length 4.5-5.5 mm. Coppery to yellowish bronze, apex of elytra with yellow marking. Antennae and legs almost all black, except tibiae and antennal segments 2-4 slightly paler.

Frons with seven to eight rather fine and slightly sinuate ridges. Pronotum with dense but not coarse punctuation except on disc. Elytral striae finely punctured, much narrower than third interval; intervals flat, all except second finely microsculptured and dull (Fig. 62); fourth interval with one (usually) or two punctures. Wings present. In open, dry, often sandy habitats; iv-ix. Widespread in England, Wales, lowland Scotland and Ireland except the north; abundant.
Similar species: See *N. biguttatus* (3).

Tribe ELAPHRINI

A small tribe characterised by the form of the front tibiae (Fig. 19), large eyes and irregular elytral sculpturing. Most are associated with damp habitats. There are about 50 world species in only three genera.

Key to genera of Elaphrini

1. Elytra with striae (Plate 26); frons unpunctured, with two deep longitudinal frontal furrows on each side joined by a transverse furrow .... 1. *Blethisa* Bonelli (p. 54)

- Elytra with irregular sculpturing, including raised shining areas, and punctured depressions (Plates 27, 28); frons with punctures and sometimes also fine striations ..... 2. *Elaphrus* Fabricius (p. 54)

1. *BLETHISA* Bonelli

This small, distinct genus has only two European species, of which one occurs in Britain and Ireland. Characterised by the prominent frontal striae, and the presence of elytral striae as well as irregular depressions. Superficially similar to the nebriine *Pelophila borealis*.

1. *Blethisa multipunctata* (**Linnaeus**) Plate 26

Length 10.5-13.5 mm. Black with a distinct coppery reflection. Appendages black. Frons with deep longitudinal and transverse grooves delimiting two raised areas on each side. Pronotum with wide side borders and deep punctured foveae inside sharp hind angles. Elytra with eight rather irregular, lightly-punctured striae and deep depressions on the third and fifth intervals. Wings present. On vegetated marshy lake shores and in fens; iv-viii. Widespread but very local in Britain and Ireland; sometimes abundant.

2. *ELAPHRUS* Fabricius

Four of the 11 European species occur in Britain (three in Ireland) but two are very scarce or local. The striae are replaced by rectangular raised, more or less shiny, areas (mirrors) and groups of punctures in round depressions. All are winged.

Key to species of *Elaphrus*

1. Length less than 7.5 mm; elytra each with a single very obvious shiny mirror one third from base (Plate 28) .... 4. *riparius* (Linnaeus)

- Length at least 8 mm; elytra each with more than one mirror, these not always as distinct 2

2. Tibiae brown, paler than rest of legs; head wider than pronotum (Plate 27) .... 1. *cupreus* Duftschmid

- Tibiae black; head not wider than pronotum 3

3. Upper surface sparsely punctured, dull and finely reticulate between punctures; elytral depressions shallow and indistinct, each with about three to seven widely spaced punctures (Fig. 70) .... 2. *lapponicus* Gyllenhal

- Upper surface densely punctured, shining between punctures; elytral depressions distinct, each containing 10-12 tightly clustered punctures (Fig. 71) .... 3. *uliginosus* Fabricius

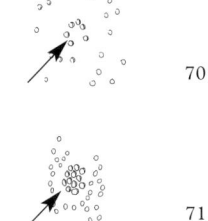

Subgenus *Elaphrus sensu stricto*

1. *Elaphrus cupreus* Duftschmid Plate 27

Length 8-9.5 mm. Coppery black with occasional purple or greenish reflections. Antennae black; tibiae brown, rest of legs black with a slight metallic, usually bluish reflection. Head wider than the quadrate pronotum; prosternum glabrous. Each elytron with at least five distinct mirrors; depressions purplish with dense punctures; intervening surfaces more sparsely punctured. In wet grassland and moorland flushes, also lowland marshes; v-vii. Widespread in Britain and Ireland; often very abundant.
Similar species: Larger than *E. riparius* (4), the only other species with pale tibiae and head wider than pronotum; coppery or brown rather than green, with more than two mirrors on each elytron.

2. *Elaphrus lapponicus* Gyllenhal

Length 8.5-11 mm. Colour variable but uniform in any one specimen; black, coppery, reddish golden, purple, blue or green. Appendages black with coloured reflection matching body colour. Head not wider than the transverse pronotum; frons with both punctures and longitudinal striations. Prosternum pubescent. Elytra markedly narrowed basally, each with many indistinct mirrors and depressions, and dull reticulate ground sculpture; punctures between depressions widely scattered; depressions with three to seven punctures (Fig. 70). In upland bogs and by moorland streams; v-vii. Very local in Cumbria, northern and central Scotland, Hebrides and Shetland; sometimes abundant.
Similar species: Flatter and with less pronounced sculpturing than all other species; duller than *E. uliginosus* (3), pronotum with sides less sinuate and hind angles not acute.

3. *Elaphrus uliginosus* **Fabricius**

Length 8-10 mm. Ground colour greenish coppery black, with brighter green areas, and purple elytral depressions; appendages black with slight green reflection. Head no wider than the convex and transverse pronotum, which has strongly sinuate sides and acute hind angles. Frons strongly punctured, hardly striated. Prosternum glabrous. Elytra little narrowed basally, densely punctured, with distinct mirrors and depressions; depressions with 10-12 punctures (Fig. 71). In oligotrophic fens and bogs; v-ix. Very local in western England, Wales and Scotland, although there are old records from south-east England; in Ireland only from Kerry; usually very scarce.
Similar species: See *E. lapponicus* (2). Head narrower and tibiae darker than *E. cupreus* (1) and *E. riparius* (4).

Subgenus *Trichelaphrus* Semenov-Tian-Shanskij

4. *Elaphrus riparius* **(Linnaeus)** Plate 28

Length 6.5-7.5 mm. Olive green (rarely brown) with coppery highlights, elytral mirrors and depressions; appendages metallic green excepting brown tibiae. Head wider than the slightly transverse pronotum, frons densely and finely punctured, the punctures elongate and confluent near the mid line. Each elytron with a single prominent mirror and numerous strongly punctured depressions. On the margins of fresh water where there is damp bare ground; v-viii. Widespread in Britain, especially in the south, and in Ireland; often abundant.
Similar species: See *E. cupreus* (1).

Tribe LORICERINI

The single world genus *Loricera* is sometimes elevated to a subfamily. There are only ten species, with one in Europe, characterised by the long antennal setae (Fig. 12). It is specialised for feeding on Collembola.

1. *Loricera pilicornis* **(Fabricius)** Plate 29

Length 6-8 mm. Metallic black or dark brown, with a greenish or coppery metallic reflection; apex and sides of elytra often paler. Antennae dark brown with basal segment black; legs brown with darker femora. Basal four antennal segments not pubescent but with long, conspicuous dark setae; segments 5-11 pubescent, some smaller setae also present on segments 5 and 6. Eyes large, vertex of head with central longitudinal depression. Pronotum with rounded sides, contracted basally to rounded hind angles. Elytra each with ten punctured striae; fourth interval with three punctures in distinct depressions. Wings present. In grasslands, damp woodland, cultivated fields, gardens and near standing or running fresh water; iv-ix. Widespread in both Britain and Ireland; very abundant.

Tribe SCARITINI

A worldwide group of subterranean carabids, characterised by the scutellum on a peduncle between the elytra and pronotum (as in Fig. 21), and fossorial front legs (Fig. 24). There are only two British genera.

Key to genera of Scaritini

1. Length at least 5 mm; outermost elytral interval punctured throughout (side view Fig. 72) .... 1. *Clivina* Latreille (p. 57)

- Length usually less than 5 mm; outermost elytral interval largely unpunctured (side view Fig. 73) .... 2. *Dyschirius* Bonelli (p. 57)

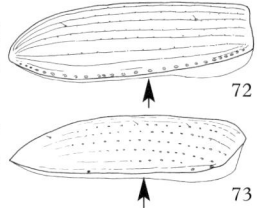

1. CLIVINA Latreille

A large genus of more than 40 species but only five in Europe and two in Britain (one in Ireland). Larger than *Dyschirius*, with many marginal punctures on the elytra (Fig. 72), a fully bordered pronotum and mid tibiae with a prominent external spine near apex.

Key to species of *Clivina*

1. Elytra light brown, paler than pronotum but often with a darker mark apically; length usually less than 6 mm .... 1. *collaris* (Herbst)

- Body unicolorous dark brown (unless immature) (Plate 30); length usually more than 6 mm .... 2. *fossor* (Linnaeus)

1. *Clivina collaris* (Herbst)

Length 5-6 mm. Body slender, almost parallel-sided. Head and pronotum dark brown, elytra pale brown, often with a darker sutural mark in apical half. Appendages brown, front femora and antennal bases sometimes darker. Wings present. In clay, sand or silt near fresh water; iv-vii. Widespread in England and Wales, local in Scotland; usually scarce.
Similar species: Usually smaller than *C. fossor* (2); colour distinct when mature but immature specimens of both species are pale brown and difficult to separate other than by size.

2. *Clivina fossor* (Linnaeus) Plate 30

Length 6-6.8 mm. Slender but less parallel-sided than *C. collaris*, elytra flattened dorsally and widened behind. Entirely unicolorous, typically dark brown, immatures pale brown. Appendages pale to mid brown. Wings often absent. In almost all open habitats, especially arable land, pasture and gardens; iv-ix. Widespread in Britain and Ireland; often very abundant.
Similar species: See *C. collaris* (1).

2. DYSCHIRIUS Bonelli

This very large genus has about 300 world species with 60 in Europe. Eleven occur in Britain but only seven of these are found in Ireland. A further British species may be currently included under *D. nitidus*, q.v. They are small cylindrical beetles, usually smaller than *Clivina*, and separated from that genus also by the reduced marginal elytral punctures, incomplete pronotal side borders and simple mid tibiae. Many are associated with the burrows of the staphylinid *Bledius* species. Body lengths given for this genus are measured from the base (rather than the tip) of the protruding mandibles. All species are winged unless stated otherwise.

Key to species of *Dyschirius*

1. Elytral striae almost absent from apical third of elytra which are short and globular (Plate 32); length less than 3 mm .... 6. *globosus* (Herbst)

- Some elytral striae distinct to near apex of the elytra, which are more elongate (Plate 31); length more than 3 mm 2

2. Base of elytra with a fine border from each shoulder to the peduncle (Fig. 74) 3

- Base of elytra not bordered 6

3. Front margin of clypeus with a median projection or tooth (Fig. 75) 4

- Front margin of clypeus straight, without a median tooth (Figs 78, 79) 5

4. Surface dull, microsculpture evident .... 3. *obscurus* (Gyllenhal)

- Surface shining, without obvious microsculpture .... 4. *thoracicus* (Rossi)

5. Length less than 4 mm; head wrinkled between eyes .... 1. *angustatus* (Ahrens)

- Length more than 4 mm; head smooth between eyes .... 2. *extensus* Putzeys

6. Front tibiae with at least one distinct sharp tooth on outer side (Fig. 76); elytra with 2-3 humeral punctures just behind shoulders 7

- Front tibiae with at most a blunt tubercle on outer side (Fig. 77); elytra with a single humeral puncture behind shoulders 9

7. Hind margin of clypeus marked by a V-shaped groove (Figs 78, 79) 8

- Hind margin of clypeus marked by a transverse groove, or indistinct .... 11. *salinus* Schaum

8. Frons with a median ridge extending backwards from the V-shaped clypeal margin (Fig. 78); base of each elytron with a small tubercle in front of scutellary pore .... 8. *luedersi* Wagner

- Frons without a median ridge (Fig. 79); base of elytra without tubercles .... 5. *aeneus* (Dejean)

9. Elytral striae deep and virtually unpunctured, intervals convex .... 7. *impunctipennis* Dawson

- Elytral striae shallow and clearly punctured (at least near base), intervals flat 10

10. Length more than 4.5 mm; elytral striae deeper basally, their punctures closer, some separated by only one or two puncture widths .... 9. *nitidus* (Dejean)

- Length less than 4.5 mm; elytral striae shallower basally, their punctures further apart, separated by at least three to four puncture widths .... 10. *politus* (Dejean)

Subgenus *Dyschirius sensu stricto*

1. *Dyschirius angustatus* (Ahrens)

Length 3.1-3.5 mm. Body very elongate and parallel-sided; dark brown with a coppery reflection. Appendages reddish brown, except antennal segments 4-11 black; front tibiae with short apical spur and two external teeth. Front of clypeus without tooth, frons between eyes with irregular wrinkles. Pronotum very slightly elongate, bordered to posterior lateral puncture. Elytra 1.9 times as long as their combined maximum width, their base bordered; striae finely but distinctly punctured. On sand and fine gravel near water, both coastal and by rivers; v-vii. Recently known only from Sussex, Cumbria and north-east Scotland; scarce.
Similar species: Smaller but relatively wider than *D. extensus* (2). Separated from *D. politus* (10), which has similarly narrow and parallel-sided elytra, by the bordered elytral bases and the front tibiae with a short apical spur and two external teeth.

2. *Dyschirius extensus* Putzeys

Length 4.1-4.9 mm. Very elongate and parallel-sided, black or dark brown with only a faint metallic reflection, elytra sometimes reddish. Appendages reddish brown; front

tibiae with long apical spur, without external teeth. Front of clypeus without tooth, frons between eyes with a transverse groove. Pronotum clearly longer than wide, sides fully bordered. Elytra more than twice as long as wide, base bordered, striae strongly punctured. On coastal sand near water; v-vii. Formerly extremely local in south-east England but not recorded since 1940.
Similar species: More slender than all other species of the genus, with pronotum clearly elongate. See also *D. angustatus* (1).

3. *Dyschirius obscurus* (Gyllenhal)

Length 3.5-4.5 mm. A very convex species, uniformly brassy black, surface dulled by fine microsculpture. Appendages dark brown or black; front tibiae toothed externally. Clypeus with median tooth. Elytra strongly convex, 1.6 times as long as wide, striae deep and almost unpunctured, intervals convex. On damp bare sand near standing water in gravel pits and lakes; v-vii. Extremely local, Sussex, Kent and Norfolk in England; Lough Neagh, Ireland; sometimes abundant.
Similar species: *D. thoracicus* (4) also has a toothed clypeus but *D. obscurus* is less shiny, with unpunctured elytral striae; *D impunctipennis* (7) has similar unpunctured striae but is shiny and without a clypeal tooth.

4. *Dyschirius thoracicus* (Rossi) Plate 31

Length 3.5-4.5 mm. Moderately convex, shining brassy black. Femora and most of antennae black or dark brown, rest of legs and base of antennae reddish; front tibiae strongly toothed. Median clypeal tooth sharp but sometimes small (Fig. 75). Elytra about 1.7 times as long as wide, base finely bordered (Fig. 74); striae shallow, very finely punctured, intervals flat. On bare coastal sand or clay; v-vii. Widespread in saltmarshes in England, local in Wales, Scotland as far north as Fife and Ireland; sometimes abundant.
Similar species: See *D. obscurus* (3). All other species of the genus lack a clypeal tooth, except sometimes *D impunctipennis* (7), which has the elytral base not bordered, and more elongate elytra with deeper, unpunctured striae.

Subgenus *Dyschiriodes* Jeannel

5. *Dyschirius aeneus* (Dejean)

Length 3.1-3.6 mm. Shining black with a metallic reflection. Appendages black or dark brown, base of antennae paler; front tibiae with one or two small but sharp outer teeth. Hind margin of clypeus angulate but without any ridge extending back onto the frons (Fig. 79). Elytra rounded, 1.6 times as long as wide, without tubercles on their sloping base; striae clearly punctured. On bare sand or mud near water; v-vii. Local in southern and eastern England, both coastal and inland; sometimes abundant.
Similar species: Differs from *D. luedersi* (8) by the lack of a median ridge behind the clypeal margin and basal tubercles on the elytra. *D. salinus* (11) has a straight or indistinct hind clypeal margin, with more elongate elytra.

6. *Dyschirius globosus* (Herbst) Plate 32

Length 2.3-2.9 mm. A short, very rounded species. Shining black or dark brown, not metallic, appendages brown. Pronotum slightly transverse, side border extremely fine, not reaching as far as hind lateral seta. Elytra only 1.5 times as long as wide, striae strongly punctured anteriorly but almost absent from apical third of elytra. Wings absent. On damp, bare peaty or sandy ground in many types of habitat; iv-ix. Widespread throughout Britain and Ireland; often very abundant.

7. *Dyschirius impunctipennis* Dawson

Length 4-5 mm. Shining brassy black, elytra sometimes reddish. Appendages black or dark brown, antennal base paler; front tibiae with long apical spur and no external teeth. Clypeal margin protruding centrally, sometimes weakly toothed. Pronotum slightly elongate. Elytra 1.8 times as long as wide, base not bordered, margin behind shoulder with a single humeral puncture, sides rounded; striae deep and unpunctured, intervals convex. On fine sand in saltmarshes; v-viii. Local in England and Wales, very local in Scotland and Ireland; sometimes abundant.
Similar species: See *D. obscurus* (3) and *D. thoracicus* (4). *D. nitidus* (9) and *D. politus* (10) have similar front tibiae but their elytral striae are clearly punctured.

8. *Dyschirius luedersi* Wagner

Length 3.3-3.9 mm. Shining brassy black. Appendages black; front tibiae with a small but sharp external tooth. Hind margin of clypeus angled and extending back medially as a fine ridge on the frons (Fig. 78). Elytra 1.6 times as long as wide, sloping base not bordered but with a small tubercle on each side; striae strongly punctured basally. On bare mud near water, both coastal and inland; v-viii. Widespread in southern and eastern England, very local in northern England, southern Scotland and Ireland; sometimes abundant.
Similar species: See *D. aeneus* (5). *D. salinus* (11) has a transverse or indistinct hind clypeal margin, and more elongate elytra.

9. *Dyschirius nitidus* (Dejean)

Length 4.6-5.8 mm. A large and convex species with rather parallel-sided elytra; brassy black, usually shining. Appendages black or dark brown; front tibiae with long apical spur and no external teeth (Fig. 77). Clypeal margin straight. Elytral base not bordered, margin behind shoulder with a single humeral puncture; striae deeply punctured basally, some punctures separated by less than twice their width; striae deep even at apex of elytra; third interval with two or three dorsal punctures. In saltmarshes on bare sand, very occasionally inland in sandpits; v-vii. Very local in England and south-west Scotland; scarce.
Similar species: See *D. impunctipennis* (7). Longer but less elongate than *D. politus* (10) with punctures of elytral striae deeper and closer together. It is possible that specimens from the extreme south east of England, usually with only two dorsal elytral punctures, may be *D. chalceus* Erichson, 1837. Those from throughout Britain, usually with three dorsal punctures are the true *nitidus*. However the true status and separation of these two species in Britain is still being resolved.

10. *Dyschirius politus* (Dejean)

Length 4-4.5 mm. Shining black with a metallic reflection. Appendages brown with front legs and antennae (except base) darkened. Elytral striae moderately punctured basally, punctures no closer than three times their width; striae somewhat fainter apically; third interval with only two dorsal punctures. On bare sand or silt, not always near water, also in saltmarshes; v-viii. Widespread but local in England, coastal in Wales, northern England and Scotland; very local in Ireland; usually scarce.
Similar species: See *D. angustatus* (1), *D. impunctipennis* (7) and *D. nitidus* (9).

11. *Dyschirius salinus* Schaum

Length 3.5-4.5 mm. Black with a brassy reflection, elytra sometimes reddish. Appendages brown or black; front tibiae with well developed external teeth (Fig. 76). Clypeus without anterior tooth, hind margin straight or indistinct, not V-shaped. Elytra slightly rounded laterally, 1.7 times as long as wide, with two or three humeral punctures behind shoulders; striae strongly punctured, intervals flat. In saltmarshes on clay or fine silt/sand banks; v-viii. Widespread on the coast of England and Wales, very local in Scotland and Ireland; sometimes abundant.
Similar species: See *D. aeneus* (5) and *D. luedersi* (8).

Tribe BROSCINI

A relict tribe, with representatives in the Holarctic and Australian regions; there are only two European genera, both of which occur in Britain and Ireland. Characterised by the scutellum on a peduncle (Fig. 21) but non-fossorial front legs.

Key to genera of Broscini

1. Length more than 15 mm .... 1. *Broscus* Panzer (p. 62)

- Length less than 10 mm .... 2. *Miscodera* Eschscholtz (p. 62)

1. *BROSCUS* Panzer

A large, cylindrical species with very faint elytral striae. There are five European species, one of which occurs in Britain and Ireland.

1. *Broscus cephalotes* (Linnaeus) Plate 33

Length 17-22 mm. Dull black, surface finely wrinkled and with reticulate microsculpture. Appendages black, apices of antennae and palpi brown; antennae with four basal segments glabrous; front tibiae very enlarged apically but without external teeth. Frons punctured between and behind eyes. Pronotum with two setae on each side, strongly contracted basally, punctured in both front and hind transverse depressions. Elytra with faint traces of very finely punctured striae. Wings present. On sandy beaches and dunes under tidal refuse, also occasionally inland in sand pits; iv-ix. Widespread around the coasts of Britain and Ireland; often very abundant.

2. *MISCODERA* Eschscholtz

Smaller, less cylindrical and more shiny than *Broscus*, with only one lateral pronotal seta. There is a single Holarctic species.

1. *Miscodera arctica* (Paykull) Plate 34

Length 6.5-8 mm. Shining bronze or metallic-black, with rather convex, globular pronotum and elytra and a distinct constriction between them. Appendages pale golden-brown; front tibiae only slightly enlarged. Head and pronotum unpunctured except for a row of punctures across the rear of the strongly rounded pronotum. Elytra with a transverse row of pits at base; with punctured striae near base but these are completely absent towards sides and behind the middle. Wings present. On dry, sandy or gravelly open, usually mossy upland habitats such as moorland, moraines and river gravels; v-viii. Widespread in northern England and central and northern Scotland, very local in south-west England, Wales and northern Ireland; usually scarce.

Tribe TRECHINI

An enormous group of small, often rather flat beetles, usually brownish or pale, sometimes with reduced eyes. Many continental European species are adapted for life in caves. Characterised typically by the lateral mandibular seta, curved frontal furrows (Fig. 26), non-reduced apical palpal segment, and usually a recurved sutural stria (Fig. 27). There are six British genera, five in Ireland, of which four were previously considered as part of the large genus *Trechus*.

Key to genera of Trechini

1. Length not more than 3 mm 2

- Length at least 3.5 mm 3

2. Elytra densely pubescent, eyes well developed (Plate 35) ....
   .... 1. *Perileptus* Schaum (p. 64)

- Elytra with at most sparse hairs and setae, eyes very small (Plate 36) .... 2. *Aepus* Samouelle (p. 64)

3. Elytra not pubescent 4

- Elytra pubescent 5

4. Raised basal border of elytra curved and not reaching scutellum (Fig. 80) .... 3. *Trechus* Clairville (p. 65)

- Raised basal border of elytra straight and reaching scutellum (Fig. 81) .... 4. *Thalassophilus* Wollaston (p. 68)

5. Pronotum shiny, not punctured or pubescent (Plate 40) ....
   .... 5. *Blemus* Dejean (p. 68)

- Pronotum with fine punctures and pubescence, at least towards rear margin (Plate 41) .... 6. *Trechoblemus* Ganglbauer (p. 69)

1. *PERILEPTUS* Schaum

A mainly tropical genus of about 50 species. The single European species is characterised by its small size, narrowed apical palpal segment and pubescent elytra with simple sutural stria.

1. *Perileptus areolatus* (Creutzer) Plate 35

Length 2.3-2.8 mm. Dark brown or black, elytra suffused with red or lighter brown medially. Antennae dark with basal segments lighter; maxillary palpi pale with sub-apical segment darkened, apical segment much narrower but hardly shorter than preceding segment; legs light brown. Head shining. Pronotum finely punctured, sides strongly contracted to sharp, protruding hind angles. Elytra flat, pubescent, sutural stria not recurved at apex. Wings present. In fine sandy shingle by rivers; v-vii. Very local in south west England, Wales, northern England, south west Scotland and south west Ireland; very scarce.

2. *AEPUS* Samouelle

Very small, pale, wingless beetles with shortened elytra, modified for life in inter-tidal crevices. There are three European species, of which two occur in Britain and Ireland.

Key to species of *Aepus*

1. Eyes very small, not protruding (Fig. 82); apices of elytra truncate, elytra almost meeting at sutural angle (Fig. 83) ....
   .... 1. *marinus* (Ström)

- Eyes larger, protruding from side of head (Fig. 84); apices of elytra lobed, with a deep incision at sutural angle (Fig 85) ....
   .... 2. *robinii* (Laboulbène)

1. *Aepus marinus* (Ström)

Length 2.1-2.8 mm. Flat, body and all appendages pale yellow or reddish brown. Eyes very small, not protruding (Fig. 82). Head as wide as pronotum, which is narrowed sharply behind. Elytra with fine but sparse pubescence as well as slightly longer setae, apices truncate, almost meeting at suture (Fig. 83). Under stones on sand in the inter-

tidal region of rocky shores; also in rock crevices; iv-ix. Widely distributed but local on rocky coasts around Britain and Ireland; sometimes abundant.
Similar species: Distinguished from *A. robinii* (2) by the truncate apices of the elytra, greatly reduced eyes and fine elytral pubescence.

2. *Aepus robinii* (Laboulbène) Plate 36

Length 2.2-2.9 mm. Entirely pale yellow-brown, eyes protruding slightly from side of head (Fig. 84). Elytra glabrous except for three or four very long setae on each, apices lobed so that the elytra do not meet apically (Fig. 85). In sand-filled inter-tidal rock crevices; iv-viii. Local around the coasts of Britain, except northern Scotland; very local in Ireland; sometimes abundant.
Similar species: See *A. marinus* (1).

3. *TRECHUS* Clairville

A very large genus, with about 200 European species, mostly found in southern and eastern Europe where many are subterranean. Distinguished from other similar sized British trechines by the non-pubescent elytra. There are seven British species, six in Ireland, although the subgenus *Epaphius* is sometimes treated as a separate genus.

Key to species of *Trechus*

1. Length more than 5 mm 2

- Length less than 5 mm 3

2. Eyes flat, hardly protruding (Fig. 86); colour pale brown, not iridescent .... 3. *fulvus* Dejean

86

- Eyes large and protruding (Fig. 87); colour mid to dark brown, elytra iridescent .... 6. *rubens* (Fabricius)

87

3. Posterior setiferous puncture in third elytral interval situated well forward, in front of recurved first stria and nearer to suture than to apex of elytra (Fig. 88) 4

- Posterior setiferous puncture situated nearer to elytral apex, within recurved first stria and about equidistant from suture and apex (Fig. 89) 5

4. Length not exceeding 4 mm; striae strongly punctured (Plate 37) .... 2. *secalis* (Paykull)

- Length at least 4.2 mm; striae almost unpunctured .... 1. *rivularis* (Gyllenhal)

5. Length more than 4.5 mm; elytra each with seven clearly punctured striae, and with a pale sub-apical spot .... 7. *subnotatus* Dejean

- Length at most 4.1 mm; elytra each with five or six hardly punctured striae, without sub-apical spot 6

6. Wings absent; elytra shorter and more rounded (Fig. 90); aedeagus in dorsal view broader and straighter with longer apex (Fig. 91) .... 4. *obtusus* Erichson

- Wings present; elytra longer and more parallel-sided (Plate 38, Fig. 92); aedeagus in dorsal view narrower and more twisted with shorter apex (Fig. 93) .... 5. *quadristriatus* (Schrank)

Subgenus *Epaphius* Samouelle

1. *Trechus rivularis* (Gyllenhal)

Length 4.2-4.8 mm. A very shining, convex species; head and pronotum dark reddish brown, elytra almost black, iridescent. Appendages light brown, segments 2-4 of antennae slightly darkened. Pronotum with sides strongly rounded, with sharp, slightly protruding tooth at hind angles; hind margin straight. Elytra with innermost three or four striae very deep and almost unpunctured, intervals strongly convex; outer striae almost completely absent; third elytral interval with posterior puncture well removed from apex. Wings

present or absent. In lowland fens and upland mires and bogs; vii-viii. Extremely local but widely dispersed in eastern and northern England, north Wales, the extreme south and north-east of Scotland and northern Ireland; usually extremely scarce.
Similar species: Larger, darker, more convex and with the posterior setiferous elytral puncture further from the elytral apex than *T. obtusus* (4) and *T. quadristriatus* (5) which have a similarly shaped pronotum.

2. *Trechus secalis* (Paykull) Plate 37

Length 3.5-4 mm. Rather convex with a distinct constriction between pronotum and elytra. Colour shining pale to mid reddish brown, appendages yellow. Sides of pronotum strongly contracted behind and sides of hind margin sinuate, so that pronotal hind angles are almost completely rounded. Elytra with four strongly punctured striae, outer striae weak or absent; posterior setiferous puncture well removed from apex (Fig. 88). Wings absent. In damp grasslands, arable fields and woodland; v-ix. Widespread in most of England but local in Wales and very local in Scotland; sometimes abundant.

Subgenus *Trechus sensu stricto*

3. *Trechus fulvus* Dejean

Length 5.1-5.7 mm. Flat, body and appendages pale reddish brown, not iridescent. Eyes small and almost completely flat (Fig. 86). Pronotum with sides sinuate, hind angles distinct and protruding, hind margin straight. Elytra with eight deep, non-punctured striae, first interval widened apically, third interval with posterior setiferous puncture situated near elytral apex. Wings absent. Near freshwater seepages on sandy and rocky coasts, under stones above high tide line and in sea caves; v-viii. Widespread but very local around the coasts of Britain and Ireland; scarce.
Similar species: As large and flat as *T. rubens* (6) but paler, with non-iridescent elytra and less protruding eyes.

4. *Trechus obtusus* Erichson

Length 3.6-4 mm. Reddish brown, head usually somewhat darker than rest of dorsal surface. Appendages yellow or pale brown. Eyes protruding, anterior puncture inside eye separated from eye by about width of puncture. Pronotum with sides strongly rounded, hind margin almost straight, hind angles rounded but with a small tooth. Elytra with rounded sides (Fig. 90) and five or six rather shallow and hardly punctured striae; posterior setiferous puncture situated close to apex. Aedeagus broad, apex produced in lateral view, straight and truncated in dorsal view (Fig. 91). Wings absent (in British and Irish specimens) so that abdominal segments are usually visible through the rather thin elytra. In fields, gardens, moorland and dry heaths; v-x. Almost ubiquitous in Britain and Ireland; often very abundant.
Similar species: Only separable from *T. quadristriatus* (5) with certainty by the aedeagus, which is longer and differently shaped apically. *T. obtusus* is always wingless, with shorter and more rounded elytra, in contrast to the winged *T. quadristriatus*. Smaller and paler than *T. subnotatus* (7) without the pale apical spots of that species. See also *T. rivularis* (1).

5. *Trechus quadristriatus* (Schrank) Plate 38

Length 3.6-4.1 mm. Reddish brown, head and centre of pronotum sometimes darker. Anterior puncture inside eye closer to the eye than width of the puncture. Pronotal hind

angles evident. Elytra only slightly rounded at sides (Fig. 92), with six or seven hardly punctured striae; posterior setiferous puncture situated close to apex (Fig. 89). Aedeagus slender with apex short in lateral view, bent and rounded in dorsal view (Fig. 93). Wings present, usually visible through the thin elytra; flies readily. In most habitats, especially agricultural fields, gardens and other disturbed, open and dry situations; vi-x. Widespread in Britain, especially near the east coast of England; local in Ireland; often very abundant.
Similar species: See *T rivularis* (1) and *T. obtusus* (4).

6. *Trechus rubens* (Fabricius)

Length 5.3-6.5 mm. Flat, head black, pronotum and elytra dark reddish brown, elytra clearly iridescent. Appendages reddish brown, some antennal segments sometimes a little darkened. Eyes large and protruding (Fig. 87). Pronotum sinuate laterally, hind angles protruding, sharp and rather raised. Elytra long, with seven punctured striae which are weaker laterally; third interval with posterior setiferous puncture close to apex. Wings present. In woodland litter (especially in coniferous plantations) and moorland moss, as well as in litter by fresh water; vi-x. Widespread but rather local in both Britain and Ireland, mainly in the north; scarce.
Similar species: See *T. fulvus* (3).

7. *Trechus subnotatus* Dejean

Length 4.5-5 mm. Dark reddish brown, elytra each with a pale apical spot, and sometimes also a paler area near shoulder. Appendages pale brown. Eyes large and protruding. Pronotum rounded laterally with hind angles marked by a small tooth. Elytra rather wide, sides rounded, with seven clearly punctured striae. Wings absent. In man-made habitats including compost, rubble and litter; v-x. A recent immigrant to Britain and Ireland, known only from Yorkshire, Devon and near Dublin; very scarce.
Similar species: See *T. obtusus* (4).

4. THALASSOPHILUS Wollaston

Small, flat trechines with a complete basal border on the elytra (Fig. 81). There are only two European species with one in Britain. It is absent from Ireland.

1. *Thalassophilus longicornis* (Sturm) Plate 39

Length 3.5-4 mm. Flat and rather parallel-sided, head black or dark brown, pronotum and elytra reddish brown. Appendages brown, antennae very long. Head almost as wide as pronotum, eyes small but protruding. Pronotum strongly constricted towards base, slightly sinuate just in front of slightly protruding hind angles. Elytra flat, parallel-sided, not pubescent, usually with five evident and slightly punctured striae; basal border straight and extending inwards to scutellum; setiferous punctures on third interval very distinct, posterior puncture close to elytral apex. Wings present. In fine shingle by streams, rivers and in gravel pits; v-vii. Very local in Wales, Kent, northern England and western Scotland; scarce.

5. BLEMUS Dejean

This genus has a single world species, previously treated as part of *Trechus* but distinguished by its pubescent elytra.

1. *Blemus discus* (Fabricius) Plate 40

Length 4.5-5.5 mm. Moderately convex but with flat elytra; reddish brown with front of head and a transverse band across elytra dark brown to black. Appendages yellow-brown, antennae exceptionally long. Eyes large and protruding. Pronotum glabrous, sides contracted basally and sinuate in front of sharp protruding hind angles. Elytra finely pubescent, especially laterally, with distinct and strongly punctured striae; recurrent stria joining apex of stria 5. Wings present. In moist habitats with silt or mud, possibly associated with mammal burrows; vii-x. Local in England, very local in Wales and Ireland; very scarce.

6. *TRECHOBLEMUS* Ganglbauer

A small genus of five species, only one European, formerly included in *Trechus*. Distinguished by both pronotum and elytra being finely pubescent.

1. *Trechoblemus micros* (Herbst) Plate 41

Length 3.8-4.5 mm. Flat, rather parallel-sided; reddish brown, head between eyes (and sometimes apical half of elytra) somewhat darker. Appendages pale brown. Eyes small, not protruding. Pronotum finely pubescent and punctured, at least towards hind margin; sides moderately contracted and slightly sinuate in front of the non-protruding hind angles. Elytra finely pubescent, with unpunctured striae; recurrent stria joining apex of stria 3. Wings present. In damp grasslands, and various habitats near water, probably in association with small mammal burrows; v-x. Widespread but local in Britain (except northern Scotland) and Ireland; usually scarce.

Tribe BEMBIDIINI

A very large assemblage (about 2000 species) of small predatory carabids, characterised by the reduced apical maxillary palpal segment (Fig. 25). As considered here, the group includes the formerly separate tribe Tachyiini. The mandibles have a lateral seta; as this is small and not easy to see it has not been used in the key to tribes. There are over 250 European species, mostly in the genus *Bembidion*. Most of the former Tachyiini are tropical or sub-tropical. All the British tachyines were previously treated in the single genus *Tachys*, now split into separate genera. Two species, *Elaphropus quadrisignatus* (Duftschmid) and *Porotachys bisulcatus* (Nicolai) are only known from single occurrences in Co. Durham in the 19th century, and have not been included in the following keys.

Key to genera of Bembidiini

1. Elytra without striae but with irregular patches of pubescence and deep punctures; eyes very large (Plate 45) .... 3. *Asaphidion* Des Gozis (p. 73)

- Elytra with striae, at least towards the front; eyes usually smaller 2

2. Apical margin of front tibia obliquely truncate before anterior margin (Fig. 94); scutellary stria absent; length usually less than 2.5 mm 3

- Apical margin of front tibia more or less straight (Fig. 95); scutellary stria present; length usually more than 2.5 mm 4

3. Basal pronotal depression unpunctured; elytra with two punctures inside recurved sutural stria (Fig. 96) ...... 1. *Tachys* Dejean (p. 71)

- Basal pronotal depression punctured (Fig. 97); elytra with a single puncture inside recurved sutural stria (Fig. 98) ....
.... 2. *Elaphropus* Motschulsky (p. 72)

4. Sutural stria recurved at apex of elytra (Fig. 99); anterior dorsal puncture on third interval behind middle of elytra ....
.... 4. *Ocys* Stephens (p. 74)

- Sutural stria usually not recurved; anterior dorsal puncture on third interval in front of middle of elytra 5

5. Third elytral interval widened, with two dull, microsculptured areas (Plate 48) .... 6. *Bracteon* Bedel (p. 75)

- Third elytral interval not widened, without dull areas 6

6. Third elytral interval with four punctures bearing long setae; head and pronotum metallic, elytra pale, surface with coarse microsculpture (Plate 47) .... 5. *Cillenus* Samouelle (p. 75)

- Third elytral interval with two fine setae; colour and sculpturing not as above .... 7. *Bembidion* Latreille (p. 76)

1. *TACHYS* Dejean

A group of hundreds of species, mostly American and many tropical. Flatter and more strongly microsculptured than *Elaphropus* species. There are four British species but none occur in Ireland.

Key to species of *Tachys*

1. Head black, rest of body uniformly dark brown to black 2

- Differently coloured, at least elytra yellow or pale brown 3

2. Antennae longer, all segments clearly elongate (Fig. 100) ....
.... 1. *bistriatus* (Duftschmid)

- Antennae shorter, middle segments in particular rounded and hardly elongate (Fig. 101) .... 3. *obtusiusculus* (Jeannel)

3. Pronotum and elytra uniformly reddish-brown, head slightly darker (Plate 42) .... 2. *micros* (Fischer von Waldheim)

- Head and pronotum dark brown or black, elytra pale except for dark scutellary mark and sometimes darker apices and sides ....
.... 4. *scutellaris* Stephens

Subgenus *Paratachys* Casey

1. *Tachys bistriatus* (Duftschmid)

Length 1.6-2.2 mm. Dark brown to black, surface strongly microsculptured. Appendages yellowish-brown, base of femora, middle of tibiae and most of antennae slightly darker; antennal segments clearly elongate, rather parallel-sided (Fig. 100). Pronotum transverse but less than 1.5 times as wide as long. Elytra rather straight-sided, with fine unpunctured striae; recurved sutural stria containing two punctures, the anterior one near the hooked apex of the stria. Wings present. On damp sand and clay near fresh water; v-vii. Extremely local in southern and eastern England; very scarce.
Similar species: With more elongate and parallel-sided elytra and longer antennae than *T. obtusiusculus* (3), the only other all dark species.

2. *Tachys micros* (Fischer von Waldheim) Plate 42

Length 2-2.4 mm. Head mid brown, pronotum and elytra pale reddish brown. Antennae brown, sometimes darker medially; legs pale yellow. Pronotum strongly contracted basally, sinuate in front of the sharp, slightly raised hind angles; basal margin angled forwards laterally. Elytral striae fine and unpunctured; anterior puncture in the recurved sutural stria close to the weakly hooked apex of the stria (Fig. 96). Wings usually present. On damp sand at the base of coastal cliffs; v-ix. Very local in England (Dorset, Hampshire, Sussex) and Wales (Lleyn Peninsula); sometimes abundant.

3. *Tachys obtusiusculus* (Jeannel)

Length 1.5-1.9 mm. Dark brown to black, coarsely microsculptured. Legs pale brown, antennae dark with base pale; antennal segments hardly elongate, sides rounded (Fig. 101). Pronotum at least 1.5 times as wide as long. Elytra with sides rounded. Wings usually absent. In *Sphagnum* tussocks at the edge of wet flushes; v-vi. Extremely local, known only from the New Forest, Hampshire; can be locally abundant.
Similar species: See *T. bistriatus* (1).

Subgenus *Tachys sensu stricto*

4. *Tachys scutellaris* Stephens Plate 43

Length 2-2.6 mm. Head and pronotum black or dark brown, elytra yellow-brown with base and sometimes apices and sides darker. Legs and antennae mid brown, with femora and apical half of antennae darkened. Sides of pronotum rounded, hardly sinuate in front of hind angles. Elytral striae very fine, impunctate; anterior puncture within the recurved sutural stria well removed from the strongly hooked apex. Wings present. On mud in salt marshes; v-viii. Very local around the coast of southern England from Norfolk to Devon; scarce.

2. ELAPHROPUS Motschulsky

Another very rich group, of mainly old-world tropical species. Only two occur in Britain but none in Ireland. More convex and shiny than *Tachys*, without evident microsculpture.

Key to species of *Elaphropus*

1. Antennae with two or more pale brown basal segments .... 1. *parvulus* (Dejean)

- Antennae with only first segment pale, second darkened or black .... 2. *walkerianus* (Sharp)

1. *Elaphropus parvulus* (Dejean) Plate 44

Length 1.8-2.2 mm. Shining black or dark brown, elytra sometimes with a paler reddish-brown mark. Antennae dark brown with basal two or three segments paler; legs yellow. Pronotum widest in front of middle, less than 1.4 times as wide as long, sides strongly constricted and sinuate behind middle; transverse basal impression with distinct punctures (Fig. 97). Elytra with sides only slightly rounded, striae deep and punctured; recurved sutural stria enclosing only one puncture (Fig. 98). Wings present. In loose sandy and gravelly soils, often associated with human disturbance; v-vii. Widely distributed but local in England as far north as Lancashire, and Wales; usually scarce.
Similar species: *E. walkerianus* (2) is more convex, with darker antennae and more transverse pronotum.

2. *Elaphropus walkerianus* (Sharp)

Length 1.8-2.1 mm. Shining black. Antennae black or dark brown with only first segment paler; legs pale brown with bases of femora darker. Pronotum more than 1.4 times as wide

as long, only slightly constricted basally and sides not sinuate; basal impression punctured. Elytral sides rounded, striae with occasional punctures. In *Sphagnum* tussocks by wet flushes; v-viii. Extremely local in the New Forest, Hampshire, also possibly Surrey and Kent; sometimes abundant.
Similar species: See *E. parvulus* (1).

3. *ASAPHIDION* Des Gozis

A small genus of over 30 species; there are ten in Europe of which four occur in Britain and three in Ireland, two of which have only been recognised relatively recently. They are diurnal predators characterised by their extremely large eyes and lack of elytral striae.

Key to species of *Asaphidion*

1. Length less than 4.7 mm; head much wider than pronotum, which has a small keel inside each hind angle 2

- Length at least 5 mm; head hardly wider than pronotum; pronotal base without keels .... 3. *pallipes* (Duftschmid)

2. Legs entirely pale; antennae pale or gradually slightly darkened from fifth segment to apex; aedeagus in side view with short, lobed apex (Fig. 102) .... 1. *curtum* (Heyden)

- Legs with at least femero-tibial joint darkened; antennae abruptly brown or black from fifth segment to apex; apex of aedeagus not lobed (Figs 103, 104) 3

3. Sub-apical segment of maxillary palpi yellow or slightly brownish dorsally; aedeagus longer than 0.85 mm, with long but broad apex (side view Fig. 103) .... 2. *flavipes* (Linnaeus)

- Sub-apical segment of maxillary palpi dark brown or black dorsally; aedeagus shorter than 0.8 mm, with long, slender apex (side view Fig. 104) .... 4. *stierlini* (Heyden)

1. *Asaphidion curtum* (Heyden)

Length 3.8-4.5 mm. Metallic bronze, strongly sculptured, with patches of pale pubescence on pronotum and elytra. All appendages pale yellow or with antennal segments 5-10 slightly and gradually darkened to mid brown. Head much wider than pronotum. Pronotum transverse, sides sharply angled at median seta and with a small keel at hind angles. Elytra widest behind middle, sides rounded. Wings present, 1.5 times elytral length. Aedeagus longer than 0.85 mm, its apex abruptly contracted and lobed (Fig. 102), major paramere with three apical setae. On open ground on heavy soils; v-ix. Probably widespread in Britain and Ireland; abundant.
Similar species: The completely pale appendages usually separate this species from both *A. flavipes* (2) and *A stierlini* (4); typically less elongate than either of those species. Aedeagus shorter apically and more abruptly contracted in side view.

2. *Asaphidion flavipes* (Linnaeus) Plate 45

Length 3.9-4.7 mm. Dark metallic bronze. All appendages yellow or pale brown, femero-tibial joints with slight metallic darkening, segments 5-10 of antennae mid to dark brown. Pronotum hardly transverse, sides rounded at median seta. Elytra widest just behind middle, sides hardly rounded. Wings present, as long as elytra. Aedeagus longer than 0.85 mm, its apex long and broad, not lobed (Fig. 103), major paramere with three apical setae. On open soils near fresh water; v-ix. Widespread in Britain, local in Ireland; sometimes abundant.
Similar species: See *A. curtum* (1). *A stierlini* (4) has even darker appendages, especially the sub-apical palpal segment.

3. *Asaphidion pallipes* (Duftschmid)

Length 5-6 mm. Reddish or purplish bronze with white and golden pubescence. Antennae dark metallic brown or black except segments 3-4 and base of segment 2 pale brown; legs brown with darker metallic reflections on front edges of femora, tibial bases and all of tarsi. Head hardly wider than pronotum, which is quadrate, angled at lateral setae and without basal keels inside the hind angles. Elytra long, evenly rounded at sides. Wings present. On bare sand near fresh water; v-ix. Very local in Britain, often coastal in the south; recorded from the north-west of Ireland in 1906 but not since; sometimes abundant.

4. *Asaphidion stierlini* (Heyden)

Length 3.8-4.5 mm. Dark bronze-black. Appendages yellow-brown with apical half of antennae, sub-apical palpal segments and femero-tibial joints almost black. Pronotum transverse with sides angled. Elytra rather parallel-sided, sides straight and hardly diverging behind. Wings present. Aedeagus shorter than 0.8 mm, its apex long and slender (Fig. 104), major paramere with four apical setae. In gardens, chalk pits and other open situations on light soils; v-viii. Probably local in southern England; scarce.
Similar species: See *A. curtum* (1) and *A. flavipes* (2).

4. OCYS Stephens

A small genus of about ten western Palaearctic species, previously considered as part of *Bembidion* but distinguished by the elytral setiferous punctures being restricted to the posterior half of the elytra, and by the strong, apically recurved first stria which forms a raised keel above the apex of stria 8 (Fig. 99). There are two species in Britain and Ireland.

Key to species of *Ocys*

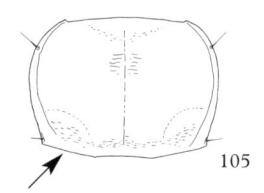

1. Pronotal base straight, hind angles sharp (Plate 46) .... 1. *harpaloides* (Audinet-Serville)

- Pronotal base curved forwards towards blunt, obtuse hind angles (Fig. 105) .... 2. *quinquestriatus* (Gyllenhal)

105

1. *Ocys harpaloides* (Audinet-Serville) Plate 46

Length 4.2-5.8 mm. A rather convex species, shining reddish brown; appendages yellow to red-brown. Frontal furrows indistinct. Pronotum broad, with wide side borders, sides sinuate just in front of the sharp, slightly raised hind angles, hind margin straight. Elytra wider than pronotum, shoulders and sides rounded, with five shallow striae that extend only two-thirds of the elytral length and width; a single dorsal puncture present in the third interval well behind middle of elytra. Wings present. In woodland under bark, and under stones in more open areas on damp soils; v-x. Widespread in Britain and Ireland except in the extreme north; very abundant.
Similar species: In addition to the feature used in the key above, can be distinguished from *O. quinquestriatus* (2) by the generally larger size, redder colour and by having only a single dorsal puncture.

2. *Ocys quinquestriatus* (Gyllenhal)

Length 3.9-4.7 mm. Dark reddish with a metallic blue reflection; appendages pale brown. Eyes rather flat. Pronotum with sides not sinuate, hind angles obtuse, hind margin angled forwards laterally (Fig. 105). Elytra with two dorsal punctures, both in apical third; striae visible almost to apex but faint or absent at sides. Wings present. Synanthropic in cellars and on walls, also on rocky coasts; vii-xii. Very local in both Britain and Ireland; very scarce.
Similar species: See *O. harpaloides* (1).

5. *CILLENUS* Samouelle

Another genus previously treated as part of *Bembidion*. There is a single very distinctive coastal species, with four dorsal setae on the characteristically coloured and sculptured elytra.

1. *Cillenus lateralis* Samouelle Plate 47

Length 3.4-4.2 mm. Very parallel-sided, with a wide head and protruding mandibles. Head and pronotum metallic greenish or coppery, elytra yellow-brown with a vague darker, slightly metallic patch apically; entire dorsal surface except frons with strong reticulate microsculpture. Appendages pale, except apical part of antennae darkened. Frontal furrows indistinct. Pronotum widest in anterior third, strongly narrowed behind and sinuate just in front of sharp hind angles. Elytra parallel-sided, with seven unpunctured striae that mostly extend to apex; third stria with four dorsal punctures bearing long setae. Wings usually absent. On the coast in saltmarshes and on open muddy sand; iv-vii. Local in Britain and Ireland; usually scarce.

6. *BRACTEON* Bedel

This genus, also known as *Chrysobracteon* Netolitzky, has also usually been considered as a subgenus of *Bembidion* but is distinct on both larval and adult features. Adults have a relatively small head (with large eyes) and pronotum, and wide, convex elytra with a widened third stria that has distinct 'silver spots' – areas of dense, reflective microsculpture, separated by smoother more shining 'mirrors', rather as in some *Elaphrus*. There are nearly 20 Holarctic species, of which only two have been found in Britain, one in Ireland. They are diurnal species that fly readily.

Key to species of *Bracteon*

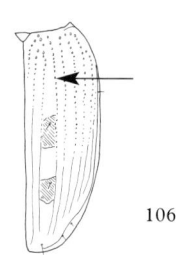

1. Elytral third and fourth striae almost straight (Fig. 106) .... 1. *argenteolum* Ahrens

- Elytral third and fourth striae abruptly curved in front of silver spots (Plate 48) .... 2. *litorale* (Olivier)

1. *Bracteon argenteolum* **Ahrens**

Length 5.8-7.3 mm. Shining dark bronze or coppery, surface strongly microsculptured. Appendages metallic black, first antennal segment and tibiae brown in part. Head shining between the frontal furrows which are strongly bent round orbital seta. Pronotum strongly transverse. Elytra with all striae more or less straight except where third stria is bent out around the silver spots (Fig. 106). On bare sand near fresh water; v-vi. Only known from Lough Neagh, Northern Ireland (where it has not been recorded since 1923), a recent single record from Kent and a single breeding site in Suffolk.

2. *Bracteon litorale* **(Olivier)** Plate 48

Length 5.6-6.2 mm. Shining bronze, coppery, reddish or purplish-metallic, microsculpture evident. Appendages entirely dark metallic. Head dull between the frontal furrows. Pronotum only slightly transverse. Third and fourth elytral striae strongly bent outwards in front of the anterior silver spot. On bare sand and fine shingle near rivers or standing water; v-vii. Very local in northern and western Britain, sometimes abundant.

7. *BEMBIDION* Latreille

This enormous genus is the largest in the family, with over 1300 species mostly in the temperate regions of the world, including about 220 in Europe. All are small, active predators or scavengers, with both diurnal and nocturnal species represented. Most usually occur near water; they are active fliers unless mentioned otherwise. The genus has been divided into many subgenera. Some of these are given generic status by various authors but there is no general agreement on the limits or sub-divisions of the genus. As considered in this work, there are 54 British species in 24 subgenera; 33 species occur in Ireland. The sometime subgenera *Bracteon*, *Ocys* and *Cillenus* are considered here as separate genera. The following key to subgenera uses superficial characters in places, and should not be used for specimens found outside Britain and Ireland. The ordering of subgenera follows the checklist rather than being alphabetical.

Several species have pale elytral markings: in order to see these, specimens should preferably be examined when live or killed and dry. Specimens preserved in liquid should have the elytra raised or detached and viewed with transmitted light.

Key to subgenera of *Bembidion*

1. Elytra each with a clearly limited pale spot behind the shoulder but anterior margin dark (Fig. 107, Plates 61, 67); often also with a second spot just behind middle but without any other spots or markings 2

 - Elytra without such well-defined two or four spots; if four-spotted, the marks are more diffuse and the anterior margin is pale (Plate 60) 3

2. Elytral striae not extending behind middle; length at least 4 mm .... 13. *Nepha* Motschulsky (p. 93)

 - Elytral striae reaching at least two-thirds of the elytral length to apex; length less than 4 mm .... 20. *Bembidion sensu stricto* (p. 97)

3. Head between and behind eyes coarsely punctured, frontal furrows replaced by punctures and wrinkles (Fig. 108) 4

 - Head not punctured, except sometimes a group of small punctures inside and behind the eyes (Fig. 109); frontal furrows more distinct 6

4. Elytra yellow with dark markings (Plate 52) .... 4. *Actedium* Motschulsky (p. 82)

 - Elytra dark metallic 5

5. Legs and antennae at least partly brown and non-metallic (Plate 51) .... 3. *Princidium* Motschulsky (p. 81)

 - Legs and antennae metallic black (Plate 53) .... 5. *Testedium* Motschulsky (p. 82)

6. Sides of pronotum rounded, hind angles not protruding (Plates 70, 71, Figs 143-145, p. 100) 7

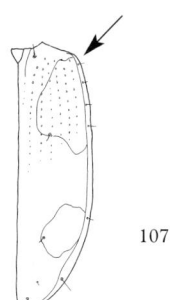

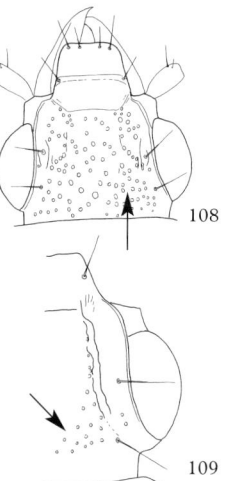

 - Sides of pronotum sinuate before hind angles, which are at least slightly protruding (as in Figs 124, 125, 133-136, p. 85, 88) 8

7. Hind margin of pronotum straight (Plate 70); elytra with a sharply raised keel from angular shoulder to base of stria 5 (Fig. 110) .... 23. *Phyla* Motschulsky (p. 99)

- Hind margin of pronotum sinuate, produced backwards medially (Plate 71, Figs 143-145, p. 100); anterior margin of elytra simple .... 24. *Philochthus* Stephens (p. 100)

8. Basal border of elytra sharply angled where it meets the lateral margin at shoulder (Fig. 111); less than 4.5 mm long, uniformly bright metallic species 9

- Basal border of elytra either absent, or meeting the lateral margin in a gradual curve (Fig. 112); length often more than 4.5 mm, if smaller, colour usually not as above 10

9. Legs black or very dark brown (Plate 49) .... 1. *Neja* Motschulsky (p. 80)

- Legs partly or entirely light brown (Plate 50) .... 2. *Metallina* Motschulsky (p. 81)

10. Frontal furrows single, deep and strongly converging in front on the clypeus (Fig. 113) 11

- Frontal furrows less distinct, sometimes doubled, never so strongly converging 12

11. Elytra unicoloured except for sub-apical spot, or with apex gradually lightened (Plate 68) ... 21. *Trepanedoris* Netolitzky (p. 98)

- Elytra distinctly patterned with pale markings (Plate 69) .... 22. *Trepanes* Motschulsky (p.98)

12. Frontal furrows doubled on clypeus (Figs 114, 115) 13

- Frontal furrows single or absent on clypeus 14

13. Frontal furrows double on clypeus but single on frons (Fig. 114) .... 17. *Semicampa* Netolitzky (p. 94)

- Frontal furrows doubled on both clypeus and frons (Fig. 115) .... 18. *Diplocampa* Bedel (p. 95)

14. Dorsal punctures on third elytral interval not touching third stria (Fig. 116) 15

- Dorsal punctures on third elytral interval touching or forming part of third stria (Fig. 117) 18

15. Head and pronotum shining metallic without microsculpture, elytra almost entirely yellowish (Plate 56) ....
   .... 8. *Notaphemphanes* Netolitzky (p. 84)

- Differently coloured; if elytra are mainly pale, head and pronotum are evidently microsculptured 16

16. Elytral striae punctured, not grooved; length less than 3.5 mm; dorsal surface mainly black, apex of elytra sometimes paler (Plate 66) .... 19. *Emphanes* Motschulsky (p. 96)

- Elytral striae grooved, with or without fine punctures; length often more than 3.5 mm; elytra patterned with pale transverse marks (Plates 54, 55) 17

17. Frontal furrows extending on to clypeus; area around anterior orbital puncture raised and shining (Fig. 118); elytra iridescent (Plate 54) .... 6. *Eupetedromus* Netolitzky (p. 82)

- Frontal furrows on frons only; frons not raised around orbital puncture; elytra not iridescent (Plate 55) ....
   .... 7. *Notaphus* Dejean (p. 83)

18. Elytral striae fainter or absent laterally and/or apically, seventh stria often faint or absent 19

- Elytra with at least seven fully developed striae, which are hardly fainter towards apex and sides 21

19. Second elytral stria well developed, almost as deep as first stria at elytral apex; femora black or dark brown, at least basally ....
   .... 11. *Bembidionetolitzkya* Strand (p. 85)

- Second elytral stria weak or absent before reaching elytral apex, where it is much shallower than first stria; femora pale yellow-brown or slightly darkened basally 20

20. Eighth stria (just inside elytral margin) complete and uninterrupted (side view Fig. 119) .... 12. *Ocydromus* Dejean (part) (p. 87)

- Eighth stria present only in apical third of elytra (side view Fig. 120) .... 14. *Sinechostictus* Motschulsky (p. 93)

21. Head with fine punctures behind and inside the eyes (Fig. 121) .... 12. *Ocydromus* Dejean (part) (p. 87)

- Head unpunctured behind and inside the eyes 22

22. Legs black or dark brown 23

- Legs at least partly paler brown or yellow 24

23. Abdominal sternites with a row of hairs along hind margins (Fig. 122) .... 10. *Trichoplataphus* Netolitzky (p. 85)

- Abdominal sternites without rows of hairs .... 9. *Plataphus* Motschulsky (p. 84)

24. Length at least 5.5 mm; eyes protruding, occupying most of side of head; base of pronotum with a deep median longitudinal furrow (Fig. 123) .... 15. *Pseudolimnaeum* Kraatz (p. 94)

- Length 4 mm or less; eyes almost flat, occupying about half side of head; base of pronotum without deep furrow (Plate 63) .... 16. *Lymnaeum* Stephens (p. 94)

1. Subgenus *Neja* Motschulsky

A western Palaearctic subgenus with ten species; one occurs in Britain but not in Ireland. Characterised by the sharp elytral shoulder angle (as in Fig. 111) and frontal furrows that are doubled behind the anterior orbital seta.

1. *Bembidion nigricorne* Gyllenhal Plate 49

Length 3.3-4 mm. Dark shining purplish bronze. Antennae black; legs black or very dark brown. Frontal furrows distinct, doubled posteriorly. Pronotum strongly contracted behind, sides barely sinuate, hind angles sharp but hardly protruding. Elytra with shoulders angled, with six punctured striae that fade out in apical third. Wings present or absent. On dry *Calluna* heaths, both upland and lowland; iii-vi, ix. Very local on lowland heaths throughout England, local in northern England, especially on the north Yorkshire moors, very local in Scotland; usually scarce.

Similar species: The dark legs and doubled frontal furrows distinguish this from *B. (Metallina) lampros* (2). The shape of the pronotal hind angles approaches that of subgenus *Philochthus* but the hind margin is straight, not protruding as in that group.

2. Subgenus *Metallina* Motschulsky

Similar to the previous subgenus, small, metallic species but with single frontal furrows. There are five or six species in both Europe and north America but only two in Britain and one in Ireland.

Key to species of *Bembidion* (*Metallina*)

1. Seventh elytral stria absent, or with not more than eight punctures, these being weaker than those of stria 6 .... 2. *lampros* (Herbst)

- Seventh elytral stria well developed, almost as strong as stria 6, with at least nine punctures .... 3. *properans* (Stephens)

2. *Bembidion lampros* (Herbst) Plate 50

Length 3-4 mm. Shining brassy or bronze, rarely bluish or green. Antennae dark with bases of first three segments paler; legs reddish brown. Frontal furrows slightly constricted in front of the anterior orbital seta. Pronotum with sides clearly sinuate in front of protruding hind angles. Elytral shoulders angled (Fig. 111), with six well developed striae and sometimes a weak trace (maximum eight shallow punctures) of stria 7. Wings present or absent. In all dry, sunny habitats, especially gardens and agricultural land; iv-ix. Ubiquitous in Britain and Ireland except in northern Scotland, where it is local; often extremely abundant.
Similar species: Distinguished from *B. properans* (3) by the lack of seventh stria and the frontal furrows. See also *B. (Neja) nigricorne* (1).

3. *Bembidion properans* (Stephens)

Length 3.5-4.4 mm. Shining brassy or bronze, sometimes bright blue or green. Antennae dark; legs brown, femora sometimes partly metallic and darkened. Frontal furrows parallel-sided in front of anterior orbital seta. Elytra with seven well-developed punctured striae, stria 7 almost as strong as 6, with at least nine punctures. Wings present or absent. The blue and green varieties are *coeruleotinctum* Reitter and *cyaneotinctum* Sharp, respectively. On dry, open clay soils; iv-viii. Widespread but local in England and Wales but only Fife and Dumfriesshire in Scotland; often scarce.
Similar species: See *B. lampros* (2).

3. Subgenus *Princidium* Motschulsky

A Palaearctic subgenus of five species, only one occurring in Britain and Ireland. Characterised by the uniformly coloured surface, punctured head (Fig. 108) and pronotum, complete striae and brown legs.

4. *Bembidion punctulatum* Drapiez Plate 51

Length 4.7-5.7 mm. Rather convex, dorsal surface entirely shining brassy or bronze-black. Head and most of pronotum with coarse punctures. Antennae black with basal segment brown; legs brown. Elytra with seven strongly punctured, fully developed striae and a long scutellary stria reaching to level with the anterior dorsal puncture on the third interval; dorsal punctures hardly deeper than those in the striae; intervals convex. Wings present. In gravel and shingle by fresh water, especially rivers and streams; v-viii. Local in northern and western Britain and in Ireland; abundant.

4. Subgenus *Actedium* Motschulsky

Characterised by the punctured metallic head and pronotum with pale, non-metallic elytra. There are two western European species but only one in Britain and Ireland.

5. *Bembidion pallidipenne* (**Illiger**) Plate 52

Length 4.3-4.9 mm. A wide and convex species; head and pronotum shining bronze, elytra pale yellow or cream with darker scutellary area and an irregular transverse dark patch behind middle. Appendages pale brown or yellow. Head with scattered large punctures. Pronotum impunctate but base wrinkled. Elytra with seven punctured striae that are absent in apical third, dorsal punctures shallow, touching stria 3. Wings present. On bare sand near fresh water, often coastal; v-vii. Very local around the coasts of both Britain and Ireland, where it is also widespread around Lough Neagh; usually scarce.

5. Subgenus *Testedium* Motschulsky

Distinguished from *Princidium* by the darker appendages, finer punctuation and incomplete striae. There are about ten species but only one occurs in Britain and Ireland.

6. *Bembidion bipunctatum* (**Linnaeus**) Plate 53

Length 3.9-4.7 mm. Black with a green, blue or brassy reflection, surface distinctly microsculptured. Appendages entirely metallic black. Head, and front and rear parts of pronotum, with coarse punctures. Elytra with seven rather finely punctured striae that fade out in the apical third, intervals flat; dorsal punctures very deep, forming two distinct foveae on each elytron. Wings present. On exposed sandy or silty deposits by both standing and running water; iv-vii. Very local in Britain and Ireland, mainly in the north and the uplands; usually scarce.

6. Subgenus *Eupetedromus* Netolitzky

There are about ten species in this subgenus, two European but only one in Britain and Ireland. Distinguished by the single frontal furrows, raised orbital puncture (Fig. 118), complete elytral striae with dorsal punctures not touching the third stria (Fig. 116), and elytral patterning and microsculpture.

7. *Bembidion dentellum* (**Thunberg**) Plate 54

Length 5.2-6 mm. Head and pronotum black with a bronze reflection, elytra iridescent bronze-brown with a pattern of paler marks, leaving the shoulders pale. Antennae black with basal segments brown; legs mid-dark brown. Frontal furrows single but deep, extending onto clypeus; anterior orbital seta surrounded by a raised shining area. Elytra with seven fully developed striae, third interval with dorsal punctures not touching the third stria. Wings present. In shaded muddy and marshy sites near water; v-viii. Widespread in England, local in Wales and Ireland, and with a single Scottish record; sometimes abundant. **Similar species:** Resembles *B. (Notaphus) varium* (10) but larger, with more iridescent elytra and different patterning.

7. Subgenus *Notaphus* Dejean

Similar elytral patterning to *Eupetedromus* but with shorter frontal furrows and simple orbital puncture. Most of the 30 species are north American. Only three are European, all of which occur in Britain but only one in Ireland. They are fully winged.

Key to species of *Bembidion* (*Notaphus*)

1. Appendages almost entirely black or very dark brown, only underside of first antennal segment sometimes paler .... 8. *obliquum* Sturm

- Legs and at least underside of basal three segments of antennae mid to pale brown 2

2. Length at most 4 mm; basal three antennal segments mostly pale, slightly darkened dorsally; elytral shoulders pale .... 9. *semipunctatum* Donovan

- Length more than 4 mm; basal three antennal segments darkened dorsally and apically; elytral shoulders dark ... 10. *varium* (Olivier)

8. *Bembidion obliquum* Sturm

Length 3.1-4.2 mm. Bronze-black, elytra with a pale pattern of two irregular transverse bands of small pale spots, shoulders and apex dark. Upper surface with fine and even reticulate microsculpture. Appendages dark, only underside of first antennal segment paler. Basal half of elytral striae finely punctured, intervals flat. On mud and near standing fresh water on acid soils; v-viii. Local in England predominantly in the east, with a single Scottish record from East Lothian; can be abundant.
Similar species: As small as *B. semipunctatum* (9) but both the elytral pattern and appendages are much darker than both that species and the larger *B. varium* (10).

9. *Bembidion semipunctatum* Donovan

Length 3.2-4 mm. Metallic bronze, elytra with two irregular transverse paler bands, as well as pale shoulders and apices. Microsculpture evident and regular on head and pronotum, irregular and finer on elytra. Antennae dark but with segments 1-3 and base of 4 predominantly pale; legs pale brown. Elytral striae deep but punctured basally, intervals slightly convex. On fine sand by rivers; v-viii. Extremely local, almost restricted to the River Severn and southern Welsh borders but also in East Anglia; very scarce.
Similar species: See *B. obliquum* (8). Resembles a smaller and paler version of *B. varium* (10) separated by smaller size, the pale elytral shoulders and paler antennal bases. The single, shallow frontal furrows distinguish it from the superficially similar *B. (Diplocampa) fumigatum* (39).

10. *Bembidion varium* (Olivier) Plate 55

Length 4.1-5 mm. Bronze or bronze-black, elytra with paler pattern and apices but shoulders dark; regular microsculpture evident. Antennae dark with at least undersides of segments 1-2 paler, segment 3 never totally pale; legs brown. Elytral striae moderately deep, punctured basally, intervals flat. On bare or partly vegetated ground near water, often in estuaries and saltmarshes; v-ix. Local in England and Wales, extremely local in southern and central Scotland and Ireland; abundant.
Similar species: See *B. obliquum* (8), *B. semipunctatum* (9) and *B. (Eupetedromus) dentellum* (7).

8. Subgenus *Notaphemphanes* Netolitzky

A single coastal species extending from north Africa and the Mediterranean to northern Europe. Distinguished from the preceding two subgenera by lack of microsculpture on the head and pronotum, and the mostly yellow elytra.

11. *Bembidion ephippium* (Marsham) Plate 56

Length 2.5-3.6 mm. Head and pronotum shining black with a slight metallic reflection, elytra yellow with a diffuse brown transverse mark behind middle. Appendages pale brown to yellow. Frontal furrows continued onto clypeus. Elytra with seven deep striae, which are punctured in front half; dorsal elytral punctures close to but not quite touching third stria. Wings present, usually visible through the elytra. In saltmarsh litter; iv-vii. Very locally distributed from Dorset around the southern and eastern English coast to Lincolnshire; scarce.

9. Subgenus *Plataphus* Motschulsky

This and the following subgenus are characterised by their complete striae, dorsal punctures touching third stria, dark legs and metallic surface. *Plataphus* has no pubescence on the abdominal sternites. Only one of the 40 boreal species of *Plataphus* occurs in Britain.

12. *Bembidion prasinum* (Duftschmid) Plate 57

Length 4.5-5.4 mm. A wide, very flat species; head and pronotum black with a dark greenish reflection, evidently microsculptured; elytra usually very dark reddish with a green reflection. Appendages black or dark brown and metallic, except first antennal segment red or brown. Sides of pronotum rather straight, hardly sinuate in front of sharp hind angles. Elytra long and parallel-sided, with seventh stria fully developed; all striae impunctate and complete to apex, dorsal punctures touching third stria. Underside of abdomen glabrous. Wings present. In shingle by running water; v-vii. Local in Wales, northern England and Scotland; sometimes abundant.
Similar species: Slightly larger on average than *B. (Trichoplataphus) virens* (13), dark green rather than bronze, and without pubescence on the underside of the abdomen. Superficially similar to *B. (Bembidionetolitzkya) tibiale* (17) but flatter, with unpunctured striae and well developed seventh stria.

10. Subgenus *Trichoplataphus* Netolitzky

A group of four European species, previously included in *Plataphus* but characterised by fine pubescence on the abdominal sternites (Fig. 122). There is a single species in Scotland.

13. *Bembidion virens* **Gyllenhal** Plate 58

Length 4.4-5.2 mm. Black with a metallic bronze reflection, and greenish highlights on frons and pronotal hind angles. Microsculpture coarser on head than on the very shining pronotum. Appendages black, shining metallic. Elytral striae complete, finely punctured basally. Abdominal sternites each with a row of fine punctures and hairs along their hind margins. Wings present. In shingle and coarse sand by lakes and rivers; vi-viii. Extremely local in northern Scotland; very scarce.
Similar species: See *B. (Plataphus) prasinum* (12).

11. Subgenus *Bembidionetolitzkya* Strand

About 50 Palaearctic species, characterised by dark femora, second stria as deep as first but seventh stria reduced. There are four species in Britain, although one is a very recent discovery, and three in Ireland. All are fully winged. Their identification can be difficult.

Key to species of *Bembidion* (*Bembidionetolitzkya*)

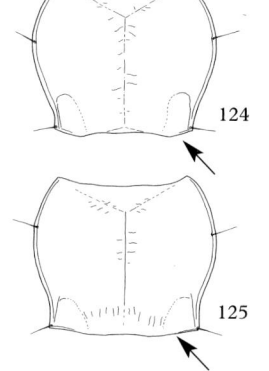

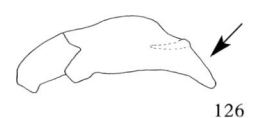

1. Basal margin of pronotum more sharply angled forwards inside hind angles (Fig. 124) 2

- Basal margin of pronotum straighter or more gradually curved forwards (Fig. 125) 3

2. Length at least 6 mm; colour bright metallic blue ....
   .... 15. *caeruleum* Audinet-Serville

- Length at most 5.5 mm; colour black with a metallic green, coppery or bluish reflection; aedeagus in side view with long, tapering and pointed apex (Fig. 126) .... 14. *atrocaeruleum* Stephens

3. Length usually at least 5.5mm; microsculpture very transverse (Fig. 127); elytral apex very blunt (Fig. 128); aedeagus in side view with short, tapering apex (Fig. 129) .... 17. *tibiale* (Duftschmid)

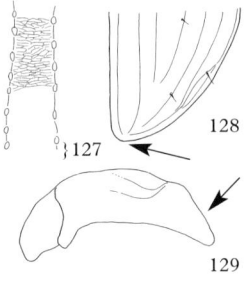

- Length usually less than 5.5mm; microsculpture less transverse; elytral apex sharper (Fig. 130); aedeagus in side view with blunt, little-tapering apex (Fig. 131) .... 16. *geniculatum* Heer

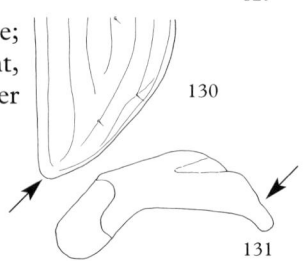

14. *Bembidion atrocaeruleum* Stephens

Length 4.5-5.5 mm. Dark metallic greenish, elytra sometimes dark brown with a greenish reflection; elytral microsculpture faint, moderately transverse. Antennae black with basal one or two segments reddish; legs mid brown with femora almost entirely dark metallic brown to black. Frontal furrows shallow, not reaching level of back of eyes. Pronotum narrow, hardly wider than head and much narrower than elytral base; basal margin angled (Fig. 124). Elytra with seven fully developed striae which are rather shallow apically; apices bluntly rounded. Aedeagus in lateral view slender, curved and with narrow, produced apex (Fig. 126). On exposed shingle by rivers and streams, mostly in the uplands; v-ix. Widely distributed in northern and western Britain, and in Ireland; sometimes abundant.

Similar species: Most similar to *B. geniculatum* (16) but with a more angled pronotal base and blunter elytral apices. Apex of aedeagus more slender than *B. geniculatum*. Smaller and more convex than *B. tibiale* (17) which has a straighter pronotal hind margin and very stout aedeagal apex.

15. *Bembidion caeruleum* Audinet-Serville

Length 6-7 mm. Metallic blue. Antennae black with basal segment and apical segments partly paler; legs dark brown with most of femora darker. Pronotum narrow, lateral part of hind margin strongly angled forwards and upwards to the hind angles. Elytra wide, somewhat ovate, with finely punctured striae and flat intervals; apices produced. On exposed sand and gravel near water; v-vii. Known only from Dungeness, Kent, first recorded in 1989; very scarce.

Similar species: Larger and brighter blue than all other species of the subgenus, with pronotal base much more angled than in *B. tibiale* (17), the only other related species of overlapping size.

16. *Bembidion geniculatum* Heer

Length 4.5-5.5 mm. Black with a slight greenish-coppery reflection; elytral microsculpture rather coarse, hardly transverse. Antennae black with basal segment brown; legs brown with femora metallic black. Frontal furrows moderately sharp,

extending almost to level with back of eyes. Pronotum at its widest almost as wide as elytral base, hind margin almost straight. Elytral striae deep to apex; apices produced (Fig. 130). Apex of aedeagus only slightly produced (Fig.131). On exposed gravel near upland streams and rivers; v-viii. Local in northern England, very local in Wales, Scotland and Ireland; usually scarce.
Similar species: See *B. atrocaeruleum* (14); smaller than *B. tibiale* (17), less green and with coarser, less transverse elytral microsculpture and more produced elytral apices. Aedeagal apex intermediate between *B. tibiale* and *B. atrocaeruleum*.

17. *Bembidion tibiale* (Duftschmid) Plate 59

Length 5.5-6.5 mm. A flat species, dark metallic green. Antennae black with first segment brown; legs brown with femora and tibial bases black. Frontal furrows deep, extending almost to level with back of eyes. Pronotal hind margin straight or only slightly curved forwards laterally (Fig. 125). Elytra long, sides only slightly rounded, microsculpture very transverse (Fig.127), striae deep and intervals convex; apices bluntly rounded (Fig. 128). Apex of aedeagus short and wide (Fig. 129). On exposed gravel and shingle by rivers and streams; v-viii. Widely distributed throughout Britain and Ireland, although extremely local in the midlands and south-east of England; very abundant.
Similar species: See *B. atrocaeruleum* (14), *B. caeruleum* (15) and *B. geniculatum* (16), also *B. (Plataphus) prasinum* (12).

12. Subgenus *Ocydromus* Dejean

This, the largest subgenus of *Bembidion*, is sometimes further split, and has also been referred to in the wider sense as *Peryphus* Dejean. They are medium sized species with single frontal furrows, pale legs, second elytral stria weak or absent apically, dorsal punctures touching the third stria, seventh striae absent or very reduced, eighth stria complete. Several have four pale spots on the elytra but these are more diffuse than the discrete marks of *Bembidion s.str.* species or *B. (Nepha) illigeri*. These markings should be looked for carefully in wet specimens by lifting the elytra and examining with the light shining through. There are at least 125 species distributed mainly across the Palaearctic region; 13 occur in Britain, ten in Ireland. They are winged unless stated otherwise.

Key to species of *Bembidion* (*Ocydromus*)

1. Head with punctures inside and just behind the eyes (Fig. 132) 2

- Head unpunctured 3

2. Length more than 5.2 mm; elytra uniformly coloured .... 20. *decorum* (Zenker in Panzer)

- Length at most 5.1 mm; elytra with at least a trace of pale markings apically, more typically with two diffuse pale spots on each .... 27. *saxatile* Gyllenhal

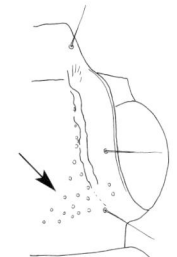

3. Head, pronotum and elytra uniformly coloured, except that elytra are sometimes slightly reddened 4

- Elytra coloured differently from head and pronotum, either uniformly brown or with pale markings (see note above) (Plate 60) 6

4. Third antennal segment and sub-apical palpal segment almost entirely pale brown; length usually more than 5.2 mm .... 28. *stephensii* Crotch

- Third antennal segment and/or sub-apical palpal segment darkened; length usually less than 5.2 mm 5

5. Pronotal base clearly punctured in middle (Fig. 133); femora usually darkened basally .... 21. *deletum* Audinet-Serville

- Base of pronotum finely striate, not or hardly punctured (Fig. 134); femora entirely pale .... 26. *monticola* Sturm

6. Elytra each with a pale shoulder mark, and a sub-apical pale area (see note above), these separated by a transverse darker band of varying width that reaches side margin (Plate 60) 7

- Elytra differently coloured 12

7. Pronotum with very narrow side borders that are absent in anterior third (as in Fig. 135) 8

- Pronotum with wider side borders that extend to front angles (as in Fig. 136) 9

8. Pronotum very convex, narrow and quadrate (Fig. 135), elytral markings distinct .... 23. *fluviatile* Dejean

- Pronotum flatter and transverse; elytral markings indistinct .... 29. *testaceum* (Duftschmid)

9. Pronotal base punctured in middle (Fig. 136) 10

- Pronotal base sometimes wrinkled but not punctured 11

10. Legs and second antennal segment pale .... 30. *tetracolum* Say

- Base of femora and second antennal segment darkened .... 18. *bruxellense* Wesmael

11. Second antennal segment more elongate, third segment completely pale (Fig. 137); pronotum usually with a metallic green reflection .... 19. *bualei* Jacquelin du Val

- Second antennal segment less elongate, third segment darkened apically (Fig. 138); pronotum shining black or with a coppery reflection .... 22. *femoratum* Sturm

12. Sides and apices of elytra completely pale (Fig. 139) .... 25. *maritimum* Stephens

- Elytra differently coloured 13

13. Elytra dark, with distinct apical pale spots but without shoulder markings (Fig. 140) .... 24. *lunatum* (Duftschmid)

- Elytra almost uniformly reddish brown, with at most a trace of apical spots 14

14. Side borders of pronotum very narrow, absent anteriorly (as Fig. 135) .... 29. *testaceum* (Duftschmid)

- Side borders of pronotum wider, present to front angles (Fig. 133) .... 21. *deletum* Audinet-Serville

18. *Bembidion bruxellense* Wesmael

Length 4.0-5.2 mm. Head and pronotum black, usually with a green reflection; elytra with pale markings rather restricted and sometimes indistinct. Antennae black with first segment and bases of segments 2-4 paler; sub-apical palpal segment black; legs pale brown, femoral bases darker. Pronotum wide and convex, wider than head, with deep basal foveae marked externally by strong ridges; centre of base strongly punctured. Elytral intervals convex dorsally, stria 7 present but very fine. On moist, usually open ground, also damp grasslands near water; v-viii. Widespread in Britain and Ireland; often abundant.
Similar species: Closest to *B. tetracolum* (30) but smaller, with darkened second and third antennal segments, as well as sub-apical palpal segment. Both have the pronotal

base punctured and a trace of the seventh elytral stria. These two features and darkened appendages also distinguish *B. bruxellense* from *B. bualei* (19) which is flatter. *B. femoratum* (22) is flatter and has a notably smaller and shinier pronotum, with its base unpunctured. The dark form of *B. saxatile* (27) is similar but flatter, with punctures inside and behind the eyes.

19. *Bembidion bualei* Jacquelin du Val

Length 4.5-5.5 mm. Head and pronotum shining black with a distinct green reflection, rarely coppery; elytra with a dark central cross and extensive pale shoulder and sub-apical marks, with at least intervals 1 and 2 dark anteriorly; legs pale brown, bases of femora sometimes darker. Antennae very long, with three pale basal segments (Fig. 137). Pronotum almost flat, base smooth. Elytral striae fine, seventh stria missing. Subspecies *anglicanum* Sharp, with darker elytral markings, is found on open fine shingle and gravel by rivers and streams mostly in the north and west of Britain; extremely local in Ireland. Subspecies *polonicum* Müller is paler and occurs on south and east coasts of England; iv-vii; usually scarce.
Similar species: Most similar to *B. femoratum* (22) but with longer and paler appendages, and greenish (rather than black or coppery) pronotum; anterior pale elytral markings usually less extensive. Smaller and flatter than *B. tetracolum* (30); see also *B. bruxellense* (18). The pale form of *B. saxatile* (27) is similar but even flatter, with punctures inside and behind the eyes.

20. *Bembidion decorum* (Zenker in Panzer)

Length 5.2-6.1 mm. Very flat, shining dark bluish green, elytra sometimes with a slight reddish tint. Antennae dark with basal segment and bases of segments 2-4 pale; legs pale brown. Head with wrinkles and a group of coarse punctures inside and behind each eye (Fig. 132); frontal furrows straight and shallow. Pronotum hardly wider than head, hind angles slightly obtuse, without ridges outside the basal foveae. Elytra elongate and parallel-sided, striae deep and closely punctured. In sand and gravel by rivers and ponds; iv-viii. Widespread in northern and western Britain and in Ireland, very local in south-east England; often abundant.
Similar species: Separated from *B. stephensii* (28), the only other uniformly greenish species of similar size and pale legs, by the punctures on the head. Larger than all other non-spotted species of *Ocydromus*, which do not exceed 5 mm. The elytra are less reddish than in *B. testaceum* (29) and it is larger than dark specimens of *B. saxatile* (27) which also have punctures inside and behind the eyes.

21. *Bembidion deletum* Audinet-Serville

Length 4.5-5.2 mm. Uniformly dark metallic green, usually with apical half of elytra slightly reddish. Antennae dark brown to black, only basal two segments completely pale; sub-apical palpal segment darkened; legs yellow to mid brown, base of femora usually darker. Pronotum rather transverse, basal foveae deep, centre of base punctured (Fig. 133). Elytra convex, striae very strongly punctured, especially near base. In damp, often shaded habitats near trickling water; v-ix. Widespread in England and Wales, very local in Scotland and Ireland; abundant.
Similar species: Smaller than *B. stephensii* (28), with darker sub-apical palpal segment (and usually also third antennal segment). More convex than the similar sized *B. monticola* (26) with the pronotum more transverse and evidently punctured basally; elytral striae also much more coarsely punctured.

22. *Bembidion femoratum* Sturm

Length 4.2-5.1 mm. Head and pronotum shining black, often with a coppery reflection; elytra with pale markings extensive, leaving only a dark central cross, with sometimes only the sutural interval dark. Antennae black, with segments 1-2 pale (Fig. 138); sub-apical palpal segment black; legs mid brown with femoral bases darker. Pronotum narrow, base sometimes with longitudinal wrinkles but without large punctures. Elytral striae fine, intervals flat, seventh stria vestigial. On open, sometimes damp, coarse sand and gravel, not always near water; iv-viii. Widespread in Britain, more local and mainly eastern in Ireland; sometimes abundant.

Similar species: See *B. bruxellense* (18) and *B. bualei* (19). The pale form of *B. saxatile* (27) is similar but flatter, with punctures inside and behind the eyes. Distinguished from *B. tetracolum* (30) by having a smaller, flatter body, a smaller and black (rather than greenish) pronotum with unpunctured base and darker appendages.

23. *Bembidion fluviatile* Dejean

Length 5.5-6.5 mm. Narrow and rather elongate and cylindrical in shape. Head and pronotum (Fig. 135) coppery or greenish, elytra with very diffuse anterior mark extending over much of the front third; sub-apical markings more distinct. Antennae very elongate, dark with basal three segments pale; sub-apical palpal segment darker brown to almost black; legs yellow to pale brown. Pronotal base usually punctured medially. Elytral striae deep, dorsal intervals convex, seventh stria well developed anteriorly. In clayish river banks; v-viii. Very local in England, Wales and southern Scotland; scarce.

Similar species: Distinguished from all other 'four-spotted' *Ocydromus* by the narrow, quadrate and very convex pronotum, with narrow side borders that do not reach the front angles. On average also larger. Narrower and more convex than *B. testaceum* (29) which has similarly narrow pronotal borders, but much less obvious elytral markings.

24. *Bembidion lunatum* (Duftschmid)

Length 5.5-6.2 mm. Body rather convex. Head and pronotum shining black with a slight green reflection, elytra dark brown with a similar reflection, and with a distinct sub-apical pale mark on each (Fig. 140). Antennae very elongate, dark with two or three pale basal segments; legs and palpi entirely pale yellow-brown. Pronotum wide and strongly contracted basally, base with fine wrinkles and some punctures. Elytral intervals slightly convex dorsally, striae much finer at sides, seventh stria absent. On silty lowland river banks, often near the coast; vii-x. Widespread but very local in England and south Wales, extremely local in Scotland and Ireland; usually scarce.

Similar species: Superficially resembles a large *B. tetracolum* (30) but can be distinguished from all other spotted *Ocydromus* by the complete absence of any pale mark on the elytral shoulders.

25. *Bembidion maritimum* Stephens

Length 5-5.5 mm. Head and pronotum dark metallic green, elytra pale cream with a central dark metallic area that is widened medially to cover five intervals on each elytron, leaving side margins completely pale (Fig. 139). Appendages pale yellow-brown, apical half of antennae sometimes slightly darkened. Pronotum with hind margin angled forwards towards hind angles, basal foveae shallow, base finely wrinkled. Elytra strongly microsculptured, with six distinct striae. On exposed fine sediments near the coast or in

estuaries; v-ix. Widespread but local around the coast of England and Wales; very local in southern Scotland and Ireland; sometimes abundant.
Similar species: Distinguished from all 'four-spotted' *Ocydromus* by the completely pale elytral sides and apices and by the very pale appendages.

26. *Bembidion monticola* Sturm

Length 4.5-5.0 mm. Uniformly metallic green or bluish. Antennae black with basal segment pale, sub-apical palpal segment slightly darkened; legs entirely pale yellow. Frontal furrows quite deep. Pronotum only slightly transverse, basal foveae not deep, with only a small ridge externally; centre of pronotal base finely striate and microsculptured but not clearly punctured (Fig. 134). Elytra rather parallel-sided; striae fairly deep but with punctures fine, almost hidden in the striae. In gravel by (usually upland) streams and rivers; v-viii. Widespread in northern England and upland Wales, very local in the rest of England, Scotland and Ireland; usually scarce.
Similar species: See *B. deletum* (21). Smaller than B. *stephensii* (28), with second and third antennal segments darkened; pronotal base with shallower foveae, only a small ridge outside these, and visible microsculpture.

27. *Bembidion saxatile* Gyllenhal

Length 4.3-5.1 mm. There are two colour forms. Both have head and pronotum shining metallic black; elytra of the commoner pale form metallic bluish brown with four distinct but diffuse, very pale, spots. Dark form with elytra dark metallic, with only a hint of paler brown markings behind shoulders and in apical half. Appendages of pale form yellow, with only apical portion of antennae darkened; dark form with all appendages mid to dark brown, most of antennae black. Punctures behind and inside the eyes very fine. Pronotal hind angles with only a hint of ridges outside the foveae. Elytra extremely flat, finely punctured striae visible almost to sides and apex. On sand and gravel near water, especially on or near the coast; v-viii. Widely distributed but very local in Britain except in northern Scotland; extremely local in Ireland; the dark form occurs inland in northern Britain; usually rather scarce.
Similar species: See *B. bruxellense* (18), *B. bualei* (19), *B. decorum* (20) and *B. femoratum* (22). It is also flatter and has more complete elytral striae than any other *Ocydromus*.

28. *Bembidion stephensii* Crotch

Length 5.3-6.1 mm. Convex and shining dark metallic green, elytra slightly reddish apically. Appendages pale yellow or reddish brown, antennal segments 4-11 darker brown. Frontal furrows very shallow. Pronotum with deep foveae and sharp ridges inside hind angles, base slightly punctured, not microsculptured. Elytra long, sides rounded and slightly widened apically, striae finely punctured. On bare clay and gravel near water; iv-vi. Local in Britain except in northern Scotland, very local in Ireland; usually scarce.
Similar species: See *B. decorum* (20), *B. deletum* (21) and *B. monticola* (26).

29. *Bembidion testaceum* (Duftschmid)

Length 4.5-5.5 mm. Rather flat; head and pronotum shining black with a strong metallic green reflection, elytra almost uniformly reddish brown with just a suggestion of a paler sub-apical mark, and a darker area around the scutellum and along the suture. Antennae dark with basal three segments pale, sub-apical palpal segment darkened; legs brown.

Pronotum quite narrow compared to elytra, side borders very narrow, not reaching front angles, hind angles hardly protruding, base not punctured. Elytra very elongate, with six well-developed striae and convex intervals, seventh stria sometimes partly developed. On sandy or gravelly river banks by slow-moving water; v-vii. Extremely local in England and south Wales; very scarce.
Similar species: See *B. decorum* (20) and *B. fluviatile* (23). Less obviously marked than any of the 'four-spotted' *Ocydromus*.

30. *Bembidion tetracolum* Say Plate 60

Length 5-6 mm. Head and pronotum dark metallic green, elytra dark brownish metallic, with paler brown diffuse shoulder mark and more distinct sub-apical markings. Antennae dark with basal three segments pale; sub-apical palpal segment pale; legs entirely pale. Pronotum wide, strongly contracted to the sharp hind angles; basal foveae deep, with a short, oblique external ridge; base with scattered strong punctures which sometimes partly merge to form deep longitudinal pits (Fig. 136). Elytra with convex intervals, seventh stria weak but distinct. Wings absent. On almost any open, not too dry soil, especially near water; iv-viii. Ubiquitous in Britain and Ireland; often very abundant.
Similar species: More convex than most other '4-spotted' *Ocydromus*; see *B. bruxellense* (18), *B. bualei* (19), *B. femoratum* (22) and *B. lunatum* (24).

13. Subgenus *Nepha* Motschulsky

A subgenus of more than a dozen Palaearctic species, only one British, characterised by distinct elytral spots, a narrow pronotum without 'double' hind angles (see *Bembidion s.str.*) and striae absent in the apical half of the elytra. A single example of a second species, *B. callosum* Küster was found in Surrey in 1851.

31. *Bembidion illigeri* Netolitsky Plate 61

Length 4.1-5.2 mm. Shining black, elytra with four pale spots, anterior margins dark (Fig. 107). Antennae black except for bases of basal three segments; legs pale yellow, apices of femora and tarsi brown. Frontal furrows single. Pronotum hardly wider than head, quadrate, contracted basally to the obtuse but sharp hind angles. Elytra with punctured striae only near base, apical half with hardly a trace of any striae. Wings present. On open sunny sites near water, sometimes coastal; v-ix. Widespread in south and eastern England, local or very local further north and west in England; only single records from Scotland and Ireland; sometimes abundant.
Similar species: Resembles *B. (s.str.) quadripustulatum* (44) but without denticulate pronotal hind angles and the striae do not extend behind the middle of the elytra.

14. Subgenus *Sinechostictus* Motschulsky

A group of about 15 mainly southern European species. The only British species is characterised by the short, apically deepened eighth stria (Fig. 120).

32. *Bembidion stomoides* Dejean Plate 62

Length 5.2-6.2 mm. A very convex species, constricted between the pronotum and elytra. Shining black, with a dark green metallic reflection, apex of elytra reddish. Appendages reddish brown, antennae darker except basal segments. Frontal furrows single, not extending on to clypeus. Pronotum hardly wider than head, strongly constricted and sinuate in front of hind angles, side borders very narrow, base strongly punctured. Elytra wide and rounded, with seven punctured striae in front two thirds; eighth stria almost absent anteriorly but strongly deepened in the apical third of elytra. Wings present. In shingle by rivers; v-vii. Local in northern England, very local in south west England, south Scotland and Wales; usually scarce.

15. Subgenus *Pseudolimnaeum* Kraatz

Of the eight Euro-Siberian species, three are European, and one has recently been discovered in Wales. They are flat, non-metallic species with complete striae, and the median pronotal line strongly deepened near hind margin.

33. *Bembidion inustum* Jacquelin du Val

Length 5.5-6 mm. An elongate flat species; head and pronotum reddish brown, elytra darker brown with only a hint of metallic reflection. Appendages pale reddish brown. Eyes large, frontal furrows single, hardly extending onto clypeus. Pronotum flat, slightly transverse, median line widened and deepened basally (Fig. 123). Elytra long and parallel-sided with shoulders rounded, and seven fully developed deep but punctured striae; dorsal punctures touching third stria. Wings present. Probably subterranean or in shingle; sometimes synanthropic in Europe. In Britain known only from three examples in mid-Wales, vi.1996 and iv.2003.

16. Subgenus *Lymnaeum* Stephens

This subgenus comprises a single, taxonomically isolated, coastal species. Characterised by the trechine-like appearance, with flat, rather parallel-sided dorsal surface and complete striae.

34. *Bembidion nigropiceum* (**Marsham**) Plate 63

Length 3.6-4.1 mm. Flat, uniformly dark to light reddish brown and moderately shining. Appendages yellow to light brown. Head as wide as pronotum; eyes small, only occupying half of side of head, which is contracted behind. Pronotum flat, strongly contracted in front of the sharp right-angled hind angles; base punctured. Elytra narrow and long-oval, widest behind middle, with eight almost unpunctured, deep striae well developed to sides and apex; dorsal punctures touching third stria, which often fuses with fourth stria at that point. Wings absent. Subterranean, in shingle and under stones on the coast; vi-viii. Very local along southern coasts of Britain, from Suffolk to south Wales; can be abundant.

17. Subgenus *Semicampa* Netolitzky

This Holarctic subgenus of about 25 species is characterised by the doubling of the frontal furrows on the clypeus but not on the frons (Fig. 114); they are small beetles without pale spots on the elytra. Only two species are found in Europe.

Key to species of *Bembidion* (*Semicampa*)

1. Upper surface unmetallic black or dark brown, legs entirely pale .... 35. *gilvipes* Sturm

- Upper surface dark metallic greenish; base of femora darkened .... 36. *schuppelii* Dejean

35. *Bembidion gilvipes* Sturm

Length 2.6-3.0 mm. Shining black or dark brown, without metallic reflection or microsculpture. Antennae black with basal three segments mostly pale; legs entirely pale brown. Pronotum strongly contracted basally, narrower at base than at front angles. Elytral striae strongly punctured at base but only stria 1 reaching elytral apex. Wings present or absent. On clay soils in shaded wet, lowland sites; xii-v. Widely distributed but local in south and east England; only two pre-1900 records from Ireland; usually scarce.
Similar species: In addition to the features in the key above, can be distinguished from *B. schuppelii* (36) by the pronotum which is narrower basally, and the lack of dorsal microsculpture. Superficially resembles *B. (Philochthus) mannerheimii* (54) in size and colour but has sinuate sides to the pronotum. Both *B. (Diplocampa) assimile* (37) and *clarkii* (38) are larger and have completely doubled frontal furrows.

36. *Bembidion schuppelii* Dejean Plate 64

Length 2.9-3.4 mm. Black with a distinct metallic green reflection; fine transverse microsculpture visible at least on elytra. Antennae dark with two basal segments paler; legs brown with basal half of femora darkened. Pronotum only moderately contracted basally, hind margin not narrower than front margin. Striae moderately punctured, at least two striae reaching elytral apices. Wings usually absent. On vegetated fine sand and gravel by streams and rivers; v-vii. Very local in northern England and southern Scotland; sometimes abundant.
Similar species: See *B. gilvipes* (35).

18. Subgenus *Diplocampa* Bedel

This Holarctic subgenus can be recognised by the frontal furrows, which are doubled on both the frons and clypeus (Fig. 115). Three of the eight species occur in both Britain and Ireland.

Key to species of *Bembidion* (*Diplocampa*)

1. Basal half of elytra with a pattern of pale and dark markings (Plate 65) .... 39. *fumigatum* (Duftschmid)

- Basal half of elytra not patterned, more or less completely dark ... 2

2. Pronotum densely microsculptured and dull ... 37. *assimile* Gyllenhal

- Pronotum smooth and very shining .... 38. *clarkii* Dawson

37. *Bembidion assimile* Gyllenhal

Length 3.0-3.5 mm. Bluish black, with strong reticulate microsculpture on head and pronotum; apex and sub-apical spot on each elytron pale brown. Antennae black or dark brown with basal segment pale; legs yellow. Elytral striae deep and strongly punctured, well developed laterally but not apically. Wings usually absent. In marshes, fens and saltmarshes; iv-vii. Local in England, Wales, south-west Scotland and Ireland; sometimes abundant.
Similar species: Smaller than *B. clarkii* (38) with a narrower and much duller pronotum, more distinct sub-apical elytral spots and usually paler elytral apices. Distinguished from *B. (Semicampa)* species (35, 36) by the fully doubled frontal furrows.

38. *Bembidion clarkii* Dawson

Length 3.3-3.8 mm. Shining black with a slight brown reflection. Head and elytra with fine microsculpture but pronotal disc smooth. Antennae black or dark brown with basal segment paler; legs brown. Pronotum very transverse, sides strongly constricted and sinuate in front of hind angles. Elytral striae fading out apically, without obvious sub-apical spots. Wings usually absent. In shaded wet sites near water, usually inland; v-viii. Local in England, very local in Wales and Ireland, only one Scottish record; sometimes abundant.
Similar species: See *B. assimile* (37) and *B. (Semicampa) gilvipes* (35).

39. *Bembidion fumigatum* (Duftschmid) Plate 65

Length 3.4-3.9 mm. Head and pronotum dark brown or black with a brassy metallic reflection, elytra with a pattern of dark and light brown markings that sometimes merge to form transverse M-shaped bands; head and pronotum densely microsculptured. Antennae dark with basal two segments pale; legs pale brown. Pronotum only moderately transverse. Elytral striae obsolete at both sides and apex, seventh stria hardly visible. Wings present. In lowland fens on clay soils, also saltmarshes; iv-vii. Widespread but local and mainly coastal in eastern England, very local in Wales; only one record from Clare in Ireland; sometimes abundant.
Similar species: Resembles *B. (Notaphus) semipunctatum* (9) in size and colouring but distinguished by having deep, doubled frontal furrows.

19. Subgenus *Emphanes* Motschulsky

This subgenus contains about 15 mostly coastal species distributed throughout the Holarctic. They are small, mostly dark coloured species with single frontal furrows; the striae are reduced apically and/or laterally and (in the British species) the dorsal punctures are separated from the third stria. Only two species occur in Britain and Ireland; both are winged.

Key to species of *Bembidion* (*Emphanes*)

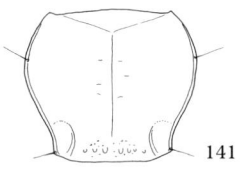

1. Legs and antennae uniformly black; pronotum strongly transverse (Plate 66) .... 40. *minimum* (Fabricius)

- Base of antennae and legs partly reddish; pronotum slightly transverse (Fig. 141) .... 41. *normannum* Dejean

141

40. *Bembidion minimum* (Fabricius) Plate 66

Length 2.4-3.3 mm. Shining black, including appendages, dorsal surface with a hint of blue reflection, apices of elytra slightly paler. Eyes large, frontal furrows deep and extending onto clypeus. Pronotum wider than head, transverse, strongly contracted basally. Elytra with seven strongly punctured striae except in apical third to quarter. Under tidal litter and in saltmarshes; v-ix. Widespread around the coast of England and Wales (rare inland) but local in Scotland and Ireland; often very abundant.
Similar species: Smaller than *B. normannum* (41), with darker appendages and a much more transverse pronotum. The pronotal shape and totally rounded elytral shoulders distinguish it from other small species, such as *B. (Phyla) obtusum* (48) and *B. (Metallina) lampros* (2).

41. *Bembidion normannum* Dejean

Length 2.7-3.6 mm. Head and pronotum shining black, elytra black to dark reddish. Antennae black with basal one or two segments partly red; legs red-brown, bases of femora darker. Pronotum rather convex dorsally, only slightly transverse (Fig. 141). Elytral striae strongly punctured until apical third or less. In tidal litter and saltmarshes; v-viii. Widespread around the coast of Britain from Humberside to south Wales, very local further north to the Solway and in southern Ireland; can be abundant.
Similar species: See *B. minimum* (40).

20. Subgenus *Bembidion sensu stricto*

There are more than ten Holarctic species in this group, with three in Britain but none in Ireland. The pronotum is strongly contracted laterally in front of the minutely toothed pronotal hind angles and incised ends of the pronotal base; the elytra have a discrete pattern of two or four pale spots leaving the front margin dark; striae present on at least the basal half of the elytra. All are winged.

Key to species of *Bembidion sensu stricto*

1. Elytra with two shoulder spots only .... 42. *humerale* Sturm

- Elytra with four spots (Plate 67) 2

2. Antennae and femora mostly pale brown or yellow ....
.... 43. *quadrimaculatum* (Linnaeus)

- Antennae and femora black ... 44. *quadripustulatum* Audinet-Serville

42. *Bembidion humerale* Sturm

Length 2.7-3.4 mm. Shining black except for one pale spot behind shoulders on each elytron. Appendages black with tibiae pale except at base. Frontal furrows single, hardly converging anteriorly. Pronotum as wide as long. Elytra with seven punctured striae which are absent in apical third. On moist, partly bare peat in lowland bogs; v-x. Known only from Hatfield and Thorne Moors in Yorkshire; sometimes abundant.

43. *Bembidion quadrimaculatum* (Linnaeus) Plate 67

Length 2.8-3.4 mm. Head and pronotum shining black with a slight metallic reflection; elytra black to mid brown with four discrete pale spots that do not reach the front margins. Appendages mid-pale brown, with apical half of antennae darker. Frontal furrows somewhat converging. Pronotum quadrate. Elytra with seven punctured striae in front half, at least the inner four striae extending well into apical half of elytra. In fields and gardens on open dry soils; iv-viii. Widespread in eastern and southern England, local in the rest of England, Wales and southern Scotland; sometimes abundant.
Similar species: Smaller than *B. quadripustulatum* (44), with much paler antennae and femora. Superficially resembles *B. (Trepanes) articulatum* (46) but in that species the whole of the front third of the elytra is pale, the frontal furrows are deeper and strongly converging and the pronotal base is straight.

44. *Bembidion quadripustulatum* Audinet-Serville

Length 3.3-4.2 mm. Shining metallic black, elytra with four discrete spots, leaving shoulders dark. Antennae black; legs pale brown with metallic dark brown or black femora. Frontal furrows only slightly convergent. Pronotum wider than long. All striae present almost to elytral apex. On damp, sometimes shaded bare mud or clay; iv-vii. Very local in southern and eastern England and east Wales; very scarce.
Similar species: See *B. quadrimaculatum* (43) and *B. (Nepha) illigeri* (31).

21. Subgenus *Trepanedoris* Netolitzky

This and the next subgenus, *Trepanes*, are characterised by the deep, convergent frontal furrows (Fig. 113); *Trepanedoris* is further distinguished by lack of pits along the pronotal hind margin and the mainly dark elytra without basal markings. There are about 15, mostly north American, species with only one in Europe.

45. *Bembidion doris* (Panzer) Plate 68

Length 3.1-3.7 mm. Shining black with a slight metallic reflection, sub-apical elytral marks and elytral apices brown. Antennae black with basal segment brown; legs reddish brown. Frontal furrows deep and converging. Head and pronotum narrow; pronotum rather cylindrical, only slightly sinuate just in front of hind angles. Elytra widest behind middle, with six or seven punctured striae well developed near base only. Wings present. In marshes and bogs; iv-vi, ix-x. Local throughout all of Britain and Ireland except the extreme north; sometimes abundant.

22. Subgenus *Trepanes* Motschulsky

This subgenus has four Palaearctic species, of which two occur in Britain but not Ireland. Characterised by the deep, convergent frontal furrows, the presence of four pits along the pronotal base, and pale markings on the elytra, including the basal half.

Key to species of *Bembidion* (*Trepanes*)

1. Basal third of elytra mostly pale (Plate 69); length more than 2.9 mm .... 46. *articulatum* (Panzer)

- Basal third of elytra dark with small elongate pale marks (Fig. 142); length less than 2.8 mm .... 47. *octomaculatum* (Goeze)

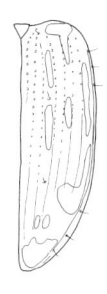

142

46. *Bembidion articulatum* (Panzer) Plate 69

Length 2.9-3.9 mm. Head and pronotum shining black with a metallic green or blue reflection; elytra mid to pale brown with scutellary region, a median transverse band and apex darkened. Antennae black with some or all of basal four segments pale brown; legs entirely pale brown. Frontal furrows very deep, almost touching at front of clypeus. Pronotum narrow especially at base, where there are four small pits between the foveate hind angles. Elytra with seven punctured striae basally, which fade before the apex. Wings present. In cracks in bare sand or mud near fresh water; vi-ix. Widespread in eastern and southern England; sometimes abundant.

Similar species: See *B. (s.str.) quadrimaculatum* (43). The deep converging frontal furrows are similar to *B. (Trepanedoris) doris* (45) but that species has the basal half of the elytra darkened.

47. *Bembidion octomaculatum* (Goeze)

Length 2.5-2.8 mm. Shining black with a slight metallic reflection; elytra with a pattern of elongate pale spots on the anterior halves of the third, fifth and seventh striae, as well as a large pre-apical spot (Fig. 142). Antennae black with only first segment completely pale; legs brown. Head with frontal furrows somewhat converging but not nearly meeting at front of clypeus. Pronotum transverse. Seven striae evident almost to apex but only punctured basally. Wings present. On bare mud near fresh water; v-vii. Extremely local in Dorset, Sussex, Kent, Berkshire and Norfolk; occasionally abundant.

23. Subgenus *Phyla* Motschulsky

A group of very small Eurasian species with sharp elytral shoulder angles and non-sinuate sides to the pronotum. There are about six species but only one occurs in Britain and Ireland.

48. *Bembidion obtusum* Audinet-Serville Plate 70

Length 2.8-3.5 mm. Dark reddish brown to black, elytral suture slightly paler. Antennae dark brown to black, basal three segments at least partly pale; legs pale brown, femoral bases darkened. Head smooth between short, single frontal furrows. Pronotum wider than head, contracted smoothly from anterior third to obtuse hind angles, hind margin straight. Elytra with sharp shoulder angles, front border continued as a line from shoulder to base of fifth stria (Fig. 110); with six shallow punctured striae that hardly reach apex or sides of elytra. Wings present or absent. On open ground and cultivated land; ii-viii. Widespread in England but increasingly local further north and west; local in Ireland; abundant.

Similar species: Distinguished from other small dark species such as *B. (Emphanes)* *minimum* (40) and *B. (Philochthus) mannerheimii* (54) by the pronotum, with its rounded sides, sharp but obtuse hind angles and straight base.

24. Subgenus *Philochthus* Stephens

About 20, mostly western Palaearctic species, characterised by the pronotum with rounded hind angles and sinuate hind margin. There are six species in Britain, four in Ireland. The only other British species likely to be confused with this subgenus are *B. (Neja) nigricorne* (1), which has hind angles only slightly protruding, and *B. (Phyla) obtusum* (48). Both have the basal margin of the elytra angled sharply at the shoulders, whereas this is rounded in *Philochthus* species.

Key to species of *Bembidion* (*Philochthus*)

1. Seventh elytral stria present as a row of distinct punctures (Plate 71) .... 50. *biguttatum* (Fabricius)

- Seventh elytral stria missing, or present as a faint line only 2

2. Legs and antennae uniformly dark brown or black .... 49. *aeneum* Germar

- Legs and sometimes base of antennae pale brown or reddish 3

3. Central part of pronotal base strongly protruding (Fig. 143); length usually more than 3.6 mm 4

- Central part of pronotal base only slightly protruding (Figs 144, 145); length usually less than 3.6 mm 5

4. Length more than 4.3 mm .... 52. *iricolor* Bedel

- Length at most 4 mm .... 53. *lunulatum* (Geoffroy in Fourcroy)

5. Pronotum less rounded at sides, narrower than the rather elongate elytra (Fig. 144); elytra with sub-apical pale spot distinct; wings usually present .... 51. *guttula* (Fabricius)

- Pronotum more rounded at sides, almost as wide as the rather short and rounded elytra (Fig. 145); apex of elytra sometimes generally paler but without distinct sub-apical spot; wings absent .... 54. *mannerheimii* Sahlberg

49. *Bembidion aeneum* Germar

Length 3.6-4.5 mm. Bronze-black, sometimes slightly green or bluish, elytra often brown towards apex and sides. Surface with distinct reticulate microsculpture. Antennae black or dark brown, basal two segments slightly paler; legs dark brown or almost black. Pronotum short and very transverse, base strongly protruding. Elytra long, widest behind middle, with six fine and only moderately punctured striae, intervals flat. Wings present or absent. On damp, poorly draining lowland soils, also coastal in estuaries and saltmarshes; v-viii. Widespread in England, Ireland, Wales and central Scotland; often abundant.
Similar species: All other species of the subgenus have paler appendages and are more shining.

50. *Bembidion biguttatum* (Fabricius) Plate 71

Length 3.9-4.4 mm. Shining black with a strong blue reflection especially on elytra, which have a distinct pale sub-apical spot. Antennae dark with basal segment pale; legs pale brown. Pronotum strongly rounded at sides, hind margin strongly sinuate. Elytra rather parallel-sided, with seven deeply punctured striae, intervals slightly convex. Wings present. On open mud and silty ground near standing fresh water; iv-vii. Widespread in England, very local in Wales and southern Scotland, absent further north and in Ireland; often very abundant.
Similar species: No other *Philochthus* has a well developed seventh stria; otherwise *B. lunulatum* (53) and *B. iricolor* (52) are similar but are somewhat smaller or larger respectively, with less developed blue reflection.

51. *Bembidion guttula* (Fabricius)

Length 2.9-3.7 mm. Shining black with at most a slight bluish or brownish reflection, elytra with pale sub-apical spot. Legs and base of antennae mid brown. Pronotum with sides moderately rounded, clearly narrower than elytra behind shoulders, hind margin abruptly but only slightly protruding medially (Fig. 144). Elytra slightly rounded at sides. Wings usually present, rarely absent. In almost all habitats, especially near water; iv-viii. Ubiquitous throughout most of Britain and Ireland; very abundant.
Similar species: Distinguished from *B. mannerheimii* (54) by the narrower pronotum in relation to the more elongate elytra; also the legs and basal antennal segment are darker, and the sub-apical elytral spot is more distinct.

52. *Bembidion iricolor* Bedel

Length 4.3-5.5 mm. Head and pronotum shining black, elytra dark brown to almost black with a slight bluish metallic reflection, apex generally paler, sub-apical spot also present. Antennae almost black with basal segments only slightly paler, very elongate, apical segments more than twice as long as wide; legs brown. Pronotum with hind margin strongly protruding. Elytra with a slight median depression near base, sides rounded; with six strongly punctured striae which fade out behind middle of elytra. Wings present. In saltmarshes and estuarine litter; iv-vi, ix-x. Local around the coast of England and Wales, only occasionally inland; usually scarce.
Similar species: Larger than *B. lunulatum* (53) with longer antennae. See also *B. biguttatum* (50).

53. *Bembidion lunulatum* (Geoffroy in Fourcroy)

Length 3.5-4 mm. Shining black with a slight brassy or brown reflection, sub-apical spot present. Antennae black with basal segments dark brown; legs brown. Pronotal sides strongly rounded and contracted basally, hind margin very strongly protruding (Fig. 143). Elytra with six strongly punctured striae, which fade out apically. Wings present. On damp bare ground near water; iv-viii. Widespread in southern England, very local in northern England and south-east Ireland; sometimes abundant.
Similar species: See *B. iricolor* (52) and *B. biguttatum* (50).

54. *Bembidion mannerheimii* Sahlberg

Length 2.8-3.4 mm. A rather short and convex species; black, only moderately shining and without metallic reflection, sub-apical spot absent. Legs and basal antennal segment pale reddish brown or yellow. Pronotum with very rounded sides, almost as wide as the rounded elytra behind shoulders (Fig. 145). Elytra with six punctured striae extending almost to apex. Wings absent. In damp grasslands and shaded habitats; v-viii. Widespread in Britain and Ireland except in the northern half of Scotland; very abundant.
Similar species: See *B. guttula* (51) and *B. (Phyla) obtusum* (48).

Tribe POGONINI

Medium sized rather parallel-sided beetles, resembling a large *Bembidion* but without the reduced apical palpal segment of that tribe. Mandibular seta and elytral basal border both distinct. There are 27 European species in three genera; the largest, *Pogonus* Dejean, with 20 European species is the only one represented in Britain (three species) and Ireland (two species). All are restricted to saltmarsh habitats and are fully winged.

Key to species of *Pogonus*

1. Upper surface entirely dark metallic 2

- Head and pronotum bright green, elytra mainly pale brown to yellow .... 3. *luridipennis* (Germar)

2. Centre of base of pronotum with punctures distinct and separate, not forming longitudinal wrinkles (Fig. 146); elytra with five punctured striae; length usually less than 7 mm .... 1. *chalceus* (Marsham)

- Centre of base of pronotum with punctures merging to form longitudinal ridges and furrows (Fig. 147); elytra with seven punctured striae; length usually at least 7 mm .... 2. *littoralis* (Duftschmid)

1. *Pogonus chalceus* (Marsham) Plate 72

Length 5.5-7 mm. Uniformly metallic coppery or greenish. Antennae black, basal segments metallic, sometimes paler; legs mid brown with darkened femoral apices. Pronotum with faint scattered punctures anteriorly, base with distinct strong punctures (Fig. 146). Elytra 1.7 times as long as together wide; with only five distinct and punctured striae; striae 6-8 very faint and unpunctured; all striae fainter apically. In saltmarshes and under tidal refuse; iv-viii. Widespread around the coasts of England and Wales, local in southern Scotland and Ireland; abundant.
Similar species: Separable from *P. littoralis* (2) not only by usually smaller size and the simply punctured pronotal base but also by having fine punctures in the front triangular pronotal impression, and by the elytra being shorter with fewer distinct striae.

2. *Pogonus littoralis* (Duftschmid)

Length 6.7-8 mm. Dark metallic coppery. Antennae black, apical segments sometimes brownish; legs brown with femora darker. Pronotum without fine punctures anteriorly, base with longitudinal wrinkles medially as well as some punctures nearer the basal foveae (Fig. 147). Elytra very long, 1.8 times as long as together wide; with all eight striae distinct, punctured and complete to elytral apices. In saltmarsh litter; v-viii. Very local around the coast of southern Britain from Lincolnshire to south Wales, and in southern Ireland; scarce.
Similar species: See *P. chalceus* (1).

3. *Pogonus luridipennis* (Germar)

Length 6-8.5 mm. Head and pronotum shining bright green with golden reflections, elytra pale brown to yellow, with a darker triangular patch behind the scutellum. Appendages pale brown to yellow, first antennal segment slightly darkened. Pronotum very transverse. Elytral striae distinct to apex. Wings present, usually visible through the elytra. On clay coastal marshes and tidal reaches under litter and seaweed; v-vii. Extremely local in Lincolnshire, Norfolk, Suffolk, the English south coast and the Severn estuary; very scarce.

Tribe PATROBINI

A small tribe of four Holarctic genera with about 20 European species. Superficially resembling large Trechini, with brown, non-metallic colouring. Characterised by the normal palpi, presence of mandibular seta (Fig. 28) but lack of basal elytral border. The three British and Irish species are in the single genus *Patrobus* Dejean, with seven European species. All are rather variable in details of punctuation, depth of striae, etc. They superficially resemble some of the smaller *Pterostichus* but are separable by their pubescent third antennal segment.

Key to species of *Patrobus*

1. Wings reduced (less than half elytral length) or absent; upper surface uniformly mid to dark reddish brown (Plate 73) 2

- Wings present; body either all black, or elytra reddish with head and pronotum darker .... 3. *septentrionis* Dejean

2. Third antennal segment about as long as basal segment (Fig. 148) .... 1. *assimilis* Chaudoir

- Third antennal segment longer than basal segment (Fig. 149) .... 2. *atrorufus* (Ström)

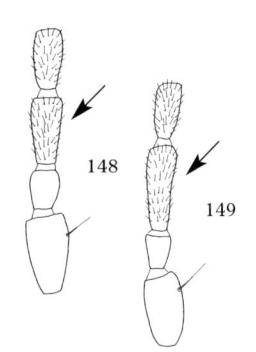

1. *Patrobus assimilis* Chaudoir

Length 6.5-9 mm. Dark reddish brown, appendages reddish, antennae only slightly darker. Antennal segments rounded, rather short, third segment not longer than first (Fig. 148). Frontal furrows parallel to side margins of head. Anterior V-shaped pronotal depression shallow but with scattered punctures both in front of and behind the line of the depression; entire pronotal base almost densely punctured. Elytra widest in apical third, sides slightly rounded; striae strong and usually punctured basally, fading out toward apex. Wings absent. On moors and in upland grasslands and woods; vi-viii. Widespread in Wales, northern England, Scotland and Ireland; abundant.
Similar species: Smaller than *P. atrorufus* (2), with shorter antennae, especially third segment. Anterior region of pronotum usually more punctured.

2. *Patrobus atrorufus* (Ström) Plate 73

Length 7.5-9.5 mm. Uniformly mid to dark reddish brown, appendages reddish. Antennal segments parallel-sided, third segment longer than the first (Fig. 149). Area between frontal furrows and side of head usually widening towards the front. Anterior pronotal depression shallow, with at most a single line of punctures along the 'V' of the depression; base with scattered punctures. Elytra widest just behind middle, sides rounded; striae often quite shallow but extending almost to apex. Wings absent. In woodlands and damp grasslands; vi-ix. Widespread in Britain except in the extreme north; very local in Ireland; abundant.
Similar species: See *P. assimilis* (1).

3. *Patrobus septentrionis* Dejean

Length 8-10 mm. Head and pronotum dark red-brown to black, elytra usually reddish, sometimes black. Antennae almost black with basal segments dark brown; legs dark reddish. Antennal segments slightly rounded, third segment just longer than first. Anterior pronotal depression often deepened medially, with a double row of punctures; basal foveae deep, punctuation of base very variable. Elytra long and flat, sides diverging to apical quarter, striae fine but continued almost to apex. Wings present. In upland habitats, usually mountain tops; v-ix. Very local in northern England, local in highland Scotland; a single Irish record from Kerry; scarce.
Similar species: More elongate than either of the other two species, with elytra longer and more parallel-sided in proportion to the forebody.

Tribe PTEROSTICHINI

A group of about 3000 species, many large and all predatory or scavenging. Previously classed with the Sphodrini and Platynini, from which they are distinguished by the more dilated apex of the front tibiae (Fig. 37) and (usually) the 'crossed' elytral epipleura (Fig. 38). Head with two punctures inside the eyes (as in Fig. 34) and labial palpi with at most two setae on the penultimate segment. Third antennal segment not pubescent. There are four British and Irish genera.

Key to genera of Pterostichini

1. First antennal segment much longer than third (Fig. 150); mandibles extremely long and produced forwards (Fig. 150, Plate 74) .... 1. *Stomis* Clairville (p. 105)

- First antennal segment about as long as third; mandibles not produced 2

2. Seventh elytral interval keeled anteriorly, behind the strong elytral shoulder tooth (Plate 87, dorso-lateral view Fig. 151) .... 4. *Abax* Bonelli (p. 115)

- Seventh elytral interval not keeled; elytral shoulder with smaller or no tooth 3

3. Antennal segments 1-3 keeled above (Fig. 152); usually brightly metallic-coloured (Plate 75) .... 2. *Poecilus* Bonelli (p. 106)

- Antennal segments 1-3 not keeled; usually black or brown, at most elytra slightly coppery in one species (Plates 76-86) .... 3. *Pterostichus* Bonelli (p. 107)

1. *STOMIS* Clairville

This genus has three European species but only one in Britain and Ireland, with characteristic antennae (Fig. 150) and mandibles.

1. *Stomis pumicatus* (Panzer) Plate 74

Length 6.5-8.5 mm. Slender and elongate, uniformly dark reddish brown to black. Appendages red-brown; basal antennal segment as long as next two segments combined. Mandibles as long as rest of head, protruding forwards prominently; frons with two deep, short furrows in front of eyes. Pronotum as long as wide, narrowed strongly to base, with single, linear foveae at hind angles. Elytra long and parallel-sided, with eight deep and punctured striae, scutellary stria and dorsal punctures absent. Wings absent. In woodlands, grasslands and disturbed ground; v-viii. Widespread in Britain and Ireland; usually scarce.

2. *POECILUS* Bonelli

A Holarctic genus of mostly metallic-coloured beetles, sometimes included within *Pterostichus*, with about 20 European species but only four in Britain, two in Ireland. Distinguished by the keeled basal antennal segments. Females are always less brightly coloured than the males.

Key to species of *Poecilus*

1. Side borders of pronotum widened posteriorly (Fig. 153); basal two antennal segments completely pale 2

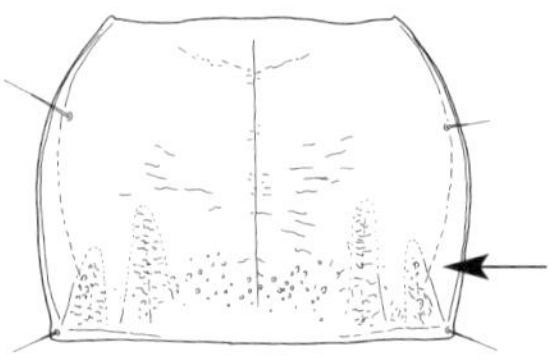

- Side borders of pronotum narrow throughout (Fig. 154); at least upper side of basal two antennal segments black or dark brown 3

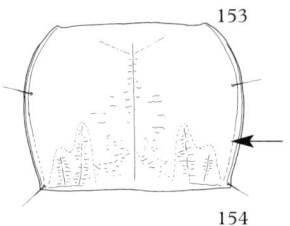

2. Head finely punctured between the eyes; hind tibiae with eight or more fine setae on inner side (Fig. 155) .... 1. *cupreus* (Linnaeus)

- Head unpunctured between the eyes; hind tibiae with five to eight coarser setae on inner side (Fig. 156) .... 4. *versicolor* (Sturm)

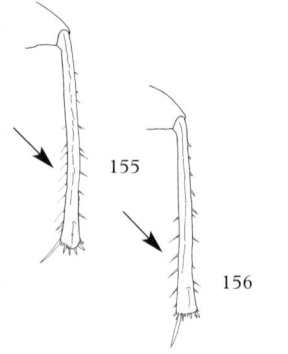

3. Basal antennal segments pale underneath; elytral striae distinctly punctured .... 2. *kugelanni* (Panzer)

- Antennae all black; elytral striae almost impunctate .... 3. *lepidus* (Leske)

1. *Poecilus cupreus* (Linnaeus) Plate 75

Length 11-13 mm. Bright coppery green, rarely bluish, purple or black. Appendages black except for pale brown basal two antennal segments. Hind tibiae with 8-10 fine setae on inner side (Fig. 155). Frons very finely punctured between the eyes. Pronotal sides evenly rounded to the slightly protruding hind angles, side borders widened posteriorly (Fig. 153); outer pronotal foveae closer to side margin than to inner fovea. Elytral striae very finely punctured. Wings present. A black form with red legs is var. *affinis* (Sturm). In dry habitats and fields; iv-vii. Widespread in southern Britain, local in northern England, southern Scotland and Ireland; can be abundant.

Similar species: Closest to *P. versicolor* (4) but distinguished by the finely punctured head and finer, more numerous tibial setae. Also larger on average, and with the outer pronotal fovea nearer to the outer margin.

2. *Poecilus kugelanni* (Panzer)

Length 12-14 mm. Typically with head and pronotum coppery red, elytra metallic green; occasionally entire upper surface coppery or black. Appendages black except paler undersides of first two antennal segments. Hind tibiae with five rather thick setae on inner side. Pronotal sides evenly rounded; side borders not widening posteriorly (Fig. 154); inner pronotal fovea deepened and extended forwards, outer foveae indistinct. Elytral striae distinctly punctured. Wings present. In sand pits, exposed sandy banks and dry heaths; iv-vii. Extremely local in southern and south-west England; usually scarce.
Similar species: Distinguished from other British *Poecilus* by the basal two antennal segments which are dark above but pale beneath.

3. *Poecilus lepidus* (Leske)

Length 11-13.5 mm. Shining green, coppery or almost black, appendages black. Hind tibiae with 6-8 extremely fine setae on inner side. Sides of pronotum slightly sinuate in front of hind angles; side borders not widened behind; foveae deep and linear. Elytral striae deep and unpunctured. Wings usually absent. On dry, open heaths; vi-ix. Very local throughout Britain; scarce.
Similar species: Separable from all other *Poecilus* species by the completely dark antennae.

4. *Poecilus versicolor* (Sturm)

Length 10.5-12.5 mm. Shining green, coppery, bluish or sometimes black. Appendages black except pale basal two antennal segments. Hind tibiae with between five and eight moderately strong internal setae (Fig. 156). Frons unpunctured. Pronotal sides evenly rounded, hind angles not protruding; side borders widened behind; outer pronotal fovea midway between outer margin and inner fovea. Elytral striae finely punctured. Wings present. In grasslands, moors and arable land, especially if wet; iv-viii. Widespread in Britain and Ireland; often abundant.
Similar species: See *P. cupreus* (1).

3. *PTEROSTICHUS* Bonelli

A very varied group with more than 200 European species, sometimes split into separate genera. There is little agreement on the names and limits of these genera, so they are treated here as subgenera. The species are generally dark coloured; they lack the basal antennal keel of *Poecilus*; the elytral margin crosses the epipleuron in all British species except *P. cristatus*. The 19 species found in Britain (15 in Ireland) include some of the commonest British carabids.

Key to species of *Pterostichus*

1. Hind angles of pronotum rounded, not toothed (Plates 77, 79) 2

- Hind angles of pronotum sharp, sometimes with a small tooth 4

2. Dorsal elytral punctures not foveate (Plate 77); apical (claw-bearing) segments of tarsi with setae beneath (as in Fig. 158) 3

- Dorsal elytral punctures foveate (Plate 79); apical segments of tarsi without setae beneath .... 5. *aterrimus* (Herbst)

3. Third elytral interval with a single dorsal puncture (behind middle); length usually at least 14 mm; femora reddish or black .... 3. *madidus* (Fabricius)

- Third elytral interval with three (rarely two) dorsal punctures; length less than 14 mm; legs always entirely black .... 2. *aethiops* (Panzer)

4. Scutellary stria absent or very faint (Plates 78, 85) 5

- Scutellary stria well developed 6

5. Tarsi furrowed above (Fig. 157); antennae black with base paler .... 17. *vernalis* (Panzer)

- Tarsi not furrowed above; antennae reddish brown .... 4. *longicollis* (Duftschmid)

6. Length more than 8 mm 7

- Length at most 7.9 mm 17

7. Claw-bearing segments of tarsi with setae beneath (Fig. 158) 8

- Claw-bearing segments of tarsi without setae beneath 9

8. Sides of pronotum regularly rounded and narrowed to a small tooth at hind angles, with a keel between basal fovea and hind angle (Plate 83) .... 11. *melanarius* (Illiger)

- Sides of pronotum sinuate in front of hind angles, which are not toothed; basal foveae deep and linear, without keel externally (Plate 76) .... 1. *cristatus* Dufour

9. Length more than 15 mm .... 7. *niger* (Schaller)

- Length at most 14 mm 10

10. Legs reddish brown; pronotum flat and strongly constricted basally (Plate 80) .... 6. *macer* (Marsham)

- Legs at last partly darkened, often black; pronotum differently shaped 11

11. Elytra with dorsal punctures foveate, often more than three on each (Plate 82) 12

- Elytra with at most three, non-foveate dorsal punctures 14

12. First antennal segment shorter than third segment (Fig. 159); pronotal base angled forwards laterally (Fig. 160); each elytron with three (rarely four, on one side) foveate punctures .... 10. *quadrifoveolatus* Letzner

- First antennal segment as long as third segment (Fig. 161); pronotal base straighter (Figs 162, 163); each elytron with 4-7 foveate punctures 13

13. Tibiae as dark as femora, usually black; side border of pronotum wider (Fig. 162) .... 8. *adstrictus* Eschscholtz

- Tibiae reddish, paler than femora; side border of pronotum extremely narrow (Fig. 163) .... 9. *oblongopunctatus* (Fabricius)

14. Side of pronotum regularly rounded and narrowed to a small tooth at hind angles (Fig. 164) 15

- Side of pronotum sinuate in front of non-toothed hind angles (Fig. 165) 16

15. Apex of male right paramere longer and more evenly rounded (Fig 166); female eighth sternite wider, less angled apically (Fig. 167) .... 15. *nigrita* (Paykull)

- Apex of male right paramere shorter and more excised (Fig. 168); female eighth sternite narrower, more sharply angled apically (Fig. 169) .... 16. *rhaeticus* Heer

16. Length at least 10 mm; abdominal sternites finely punctured .... 12. *anthracinus* (Panzer)

- Length less than 10 mm; abdominal sternites not punctured .... 13. *gracilis* (Dejean)

17. Basal foveae of pronotum doubled, outer fovea short and keeled externally (Fig. 170) .... 14. *minor* (Gyllenhal)

- Basal foveae of pronotum simple, not keeled externally (Fig. 171) 18

18. Prosternum shining and punctured above base of front legs (side view Fig. 172); legs pale brown or red .... 19. *strenuus* (Panzer)

- Prosternum dull and unpunctured; legs dark brown or black .... 18. *diligens* (Sturm)

Subgenus *Pterostichus sensu stricto*

1. *Pterostichus cristatus* (Dufour) Plate 76

Length 14-17 mm. Rather flat, shining black, elytra iridescent. Appendages black, palpi and tips of antennae lighter. Sides of pronotum strongly contracted and sinuate in front of sharp hind angles, foveae deep, unpunctured and linear, curved outwards. Elytral striae deep, intervals concave, third interval with three or four strong dorsal punctures. Elytral margin simple, not crossing epipleuron near apex. Wings absent. In woodland, gardens and shaded grassland; vi-ix. Widespread in north-east England, very local in north-west England and Scotland; sometimes abundant.
Similar species: *P. melanarius* (11) is of similar size but is separated by having a differently shaped pronotum, non-iridescent elytra and the elytral margins crossing the epipleura. *P. niger* (7) is larger, with no setae underneath the tarsal claw joints.

Subgenus *Steropus* Stephens

2. *Pterostichus aethiops* (Panzer)

Length 11.5-13.5 mm. Rather convex, body and appendages shining black. Tarsal claw joints with setae beneath. Pronotal hind angles completely rounded, basal foveae rather shallow, not clearly delimited and without external keel. Elytra widest in apical third, striae deep, third interval with three dorsal punctures. Last abdominal segment of males with a transverse tooth on the sub-apical ventral segment. Wings absent. On moorland and in upland woodlands, especially conifer plantations; v-viii. Local in south-west and northern England, Wales and Scotland except the extreme north; a single Irish record from Waterford; sometimes abundant.
Similar species: Resembles a small dark-legged *P. madidus* (3) but separated by having three dorsal elytral punctures and lacking an external keel between the pronotal foveae and side margin. Males have a tooth on the sub-apical, rather than apical ventral segment as in *P. madidus*.

3. *Pterostichus madidus* (Fabricius) Plate 77

Length 14-18 mm. Shining black, appendages black, except palpi and antennal apex paler; often with femora reddish brown (form *concinnus* (Sturm)). Tarsal claw joints with setae beneath. Pronotal hind angles rounded, basal foveae deep, delimited by a keel externally. Elytra evenly rounded, widest near middle; striae rather fine, third interval with a single (rarely two) dorsal puncture behind the middle. Males with a strong tooth on the apical ventral segment. Wings absent. In woodlands, gardens and dry grasslands; vii-ix. Ubiquitous in Britain and Ireland; often extremely abundant.
Similar species: See *P. aethiops* (2). The rounded pronotal hind angles (and often the reddened femora) distinguish it from the equally abundant *P. melanarius* (11) with which it often co-exists.

Subgenus *Pedius* Motschulsky

4. *Pterostichus longicollis* (Duftschmid) Plate 78

Length 5-6 mm. Body narrow and parallel-sided. Mid to dark reddish brown, all appendages light brown; tarsi not furrowed above. Pronotal sides sinuate in front of sharp hind angles, base punctured, basal foveae deep and linear, not doubled. Elytra without

scutellary stria, third interval with a single dorsal puncture near apex; striae punctured basally, deepened towards apex. Wings absent. On bare or sparsely vegetated ground near water; iii-v. Very local in south and east England; scarce.
Similar species: Only *P. vernalis* (17) also lacks a scutellary stria but that species is larger, darker and has a dorsal furrow on its tarsi. The brown colour and pale legs also separate *P. longicollis* from all other small species of the genus.

Subgenus *Lyperosomus* Motschulsky

5. *Pterostichus aterrimus* (Herbst) Plate 79

Length 13-15 mm. Body and appendages shining black. Tarsal claw joints without ventral setae. Head with deep, curved frontal furrows anteriorly. Pronotal sides rather straight and converging to rounded hind angles; foveae deep and well delimited externally. Elytra parallel-sided, striae rather finely punctured, outer striae very faint; third interval with three or four deep foveate punctures. Wings present. In wet humus in fens and bogs; iii-vi. Formerly extremely local in East Anglia and the New Forest, Hampshire; not recorded from Britain since 1973; widespread but very local in Ireland; scarce.
Similar species: Superficially resembles a small *P. madidus* (3), or large *P. aethiops* (2) but the pronotal and elytral sides are straighter, and the elytra have three deeply foveate dorsal punctures.

Subgenus *Adelosia* Stephens

6. *Pterostichus macer* (Marsham) Plate 80

Length 10.5-14 mm. Very flat, dark brown to black; antennae brown, legs reddish. Pronotal sides strongly contracted towards base, sinuate in front of sharp hind angles. Foveae deep, inner fovea linear, outer fovea suggested or lacking. Elytra long and parallel-sided, scutellary stria present, third interval with three dorsal punctures, the posterior one close to elytral apex. Wings present. On clay soils, often in cracks in the ground, also under bark and in coastal marshes; iv-viii. Local in eastern and southern England, and on the Welsh coast; usually scarce.

Subgenus *Platysma* Bonelli

7. *Pterostichus niger* (Schaller) Plate 81

Length 16-21 mm. Body and appendages entirely black. Tarsal claw joints without ventral setae. Pronotum rather flat, contracted gradually and slightly sinuate in front of sharp but not-toothed hind angles; foveae deep and double, inner fovea extending forwards. Elytra long and rather parallel-sided, with deep unpunctured striae and strongly raised convex intervals. Wings present. In woodland, damp grasslands and moorland; vii-ix. Widespread in Britain and Ireland; often abundant.
Similar species: See *P. cristatus* (1). Larger also than *P. melanarius* (11), without setae underneath the tarsal claw joints, and with the pronotal sides much straighter in front of the non-toothed hind angles.

Subgenus *Bothriopterus* Chaudoir

8. *Pterostichus adstrictus* Eschscholtz Plate 82

Length 10-13 mm. Black, females with elytra strongly microsculptured and often dark bronze. Appendages entirely black; first antennal segment almost as long as third (Fig. 161); claw joints of tarsi without setae beneath. Pronotal side border widened posteriorly, basal foveae often deep and linear, sometimes punctured or wrinkled; hind margin almost straight laterally (Fig. 162). Elytra with striae rather fine and unpunctured, third interval with four to six foveate punctures. Wings present. On upland moors and heaths; v-viii. Widespread in uplands of Wales, northern England, Scotland except the extreme north and Ireland except the south; abundant.

Similar species: Distinguished from *P. nigrita* (15) and related species by the deeper, foveate dorsal punctures on the elytra. Most similar to *P. oblongopunctatus* (9) but the legs are unicoloured, usually black, and the pronotal side border is wider. The first antennal segment is longer than in *P. quadrifoveolatus* (10) and the pronotal base straighter.

9. *Pterostichus oblongopunctatus* (Fabricius)

Length 9.5-12.5 mm. Shining black or slightly metallic, female elytra only slightly duller. Appendages black, tibiae and tarsi dark reddish brown. Pronotal side border extremely narrow, hardly widened posteriorly; foveae linear, not or hardly punctured on each side; hind margin almost straight (Fig. 163). Elytra with four to seven foveate dorsal punctures. Wings present. In open woodlands and dry forest; v-viii. Local in both Britain and Ireland; usually scarce.

Similar species: See *P. adstrictus* (8); separated from *P. quadrifoveolatus* (10) by the longer first antennal segment, straighter pronotal base and (usually) more than three dorsal elytral punctures.

10. *Pterostichus quadrifoveolatus* Letzner

Length 9.5-11.5 mm. Shining black, females not duller. Appendages black, first antennal segment notably shorter (and sometimes wider) than the third segment (Fig. 159). Pronotal side border slightly widened posteriorly, foveae deep, linear and usually punctured on each side; hind margin strongly angled forwards laterally (Fig. 160). Elytral striae quite strong, sometimes punctured, third interval with three (exceptionally four, on one side) foveate punctures. Wings present. In woodlands, and on lowland heaths; v-ix. A recent arrival in Britain, now distributed very locally throughout eastern England, south Wales and south-east Scotland; scarce.

Similar species: See *P. adstrictus* (8) and *P. oblongopunctatus* (9).

Subgenus *Omaseus* Stephens

11. *Pterostichus melanarius* (Illiger) Plate 83

Length 13-17 mm. Body and appendages entirely shining black. Tarsal claw joints with ventral setae (Fig. 158). Pronotum almost evenly rounded laterally, hind angles sharp with a small protruding tooth; foveae deep and wide, not clearly doubled. Elytral striae deep, intervals moderately convex. Wings usually absent. In gardens, grassland and especially agricultural fields; vi-ix. Widespread in both Britain and Ireland; very abundant.

Similar species: See *P. cristatus* (1), *P. madidus* (3) and *P. niger* (7). Much larger than *P. nigrita* (15) and *P. rhaeticus* (16) which have a similar pronotal shape.

Subgenus *Pseudomaseus* Chaudoir

12. *Pterostichus anthracinus* (Panzer)

Length 10-13 mm. Black, appendages black or dark brown. Tarsal claw joints without ventral setae. Pronotal sides gradually sinuate, hind angles sharp but not toothed (Fig. 165); foveae deep and punctured, with a keel externally. Elytra long and rather flat, with a small apical tooth in females. Ventral abdominal segments punctured. Wings present or absent. In marshes, fens and near fresh water; v-viii. Local in England, Wales, Ireland and south-west Scotland; usually scarce.
Similar species: Slightly larger than *P. nigrita* (15) and *P. rhaeticus* (16) and with the pronotal sides sinuate; separated also from *P. gracilis* (13) by larger size and the punctured underside of the abdomen.

13. *Pterostichus gracilis* (Dejean)

Length 8.5-9.5 mm. Black, appendages sometimes slightly lighter. Pronotal sides gradually sinuate, hind angles sharp, sometimes with a minute tooth; foveae punctured, rather flat with an external keel. Elytra parallel-sided but convex, apex without teeth. Abdominal sternites unpunctured. Wings present. In wet, vegetated sites near water; v-viii. Very local in England, Wales and Ireland; very scarce.
Similar species: See *P. anthracinus* (12); more slender than *P. nigrita* (15) and *P. rhaeticus* (16) and with a more sinuate pronotal side margin. Larger than *P. minor* (14) with the inner pronotal foveae not extended as far forwards.

14. *Pterostichus minor* (Gyllenhal)

Length 6.8-7.9 mm. Black; antennae black with basal one to three segments at least partly paler; legs dark reddish brown. Pronotal margins strongly sinuate, hind angles sharp, not toothed; basal foveae punctured with a small external keel, inner foveae linear and extending forwards well in front of outer foveae (Fig. 170). Elytra relatively short, not toothed apically. Wings variable, often absent. In marshes and wet grasslands; iv-viii. Widespread in England, Wales and Ireland; local in Scotland; often abundant.
Similar species: See *P. gracilis* (13). Larger than *P. diligens* (18) and *P. strenuus* (19) which do not have the pronotal foveae doubled.

15. *Pterostichus nigrita* (Paykull) Plate 84

Length 9-12 mm. Black, appendages usually also black; var. *rufifemoratus* Stephens has yellow/brown femora. Pronotum rather convex, sides strongly rounded, not sinuate in front of the toothed hind angles; basal foveae deep, punctured and with a keel externally (Fig. 164). Elytra rather convex, parallel-sided, not toothed apically. Apex of male right paramere long and evenly rounded apically (Fig. 166). Female eighth sternite relatively wide, apical angle moderately acute (Fig. 167). Wings present. In almost all damp lowland habitats, especially near fresh water; iv-viii. Widespread in Britain and Ireland; var *rufifemoratus* in Ireland only; very abundant.
Similar species: Separable from *P. rhaeticus* (16) only by the male and female genitalia; genetic studies show that even these do not guarantee complete distinction between the two species. See also *P. melanarius* (11), *P. anthracinus* (12) and *P. gracilis* (13)

16. *Pterostichus rhaeticus* **Heer**

Length 9-11.5 mm. Externally as in *P. nigrita*. Apex of male right paramere short, margin excised and sinuate (Fig. 168). Female eighth sternite narrow, apical angle very acute (Fig. 169). In wet grasslands and marshes, especially at higher altitudes but its habitats overlap with those of *P. nigrita*; iv-viii. Widespread in Britain and Ireland; very abundant.
Similar species: See *P. melanarius* (11), *P. anthracinus* (12), *P. gracilis* (13) and *P. nigrita* (15).

Subgenus *Lagarus* Chaudoir

17. *Pterostichus vernalis* **(Panzer)** Plate 85

Length 6-7.5 mm. Black, appendages dark brown to black, with basal antennal segment, tibiae and tarsi paler; tarsi furrowed above (Fig. 157). Pronotal sides evenly rounded in front of toothed hind angles; base punctured, foveae single, linear. Elytra without scutellary striae; third interval with three dorsal punctures; striae deep, hardly punctured. Wings present or absent. In most damp or shaded lowland habitats, especially grasslands; iv-viii. Widespread in Britain, except in northern Scotland, and in Ireland; abundant.
Similar species: See *P. longicollis* (4). Separated from the other small species, *P. diligens* (18) and *P. strenuus* (19) by the lack of any scutellary stria, and by the dorsally-furrowed tarsi.

Subgenus *Argutor* Stephens

18. *Pterostichus diligens* **(Sturm)**

Length 5.5-7 mm. Black, appendages dark brown or black. Pronotal sides contracted to just in front of the hind angles then slightly sinuate; base sparsely punctured, foveae single, linear. Prosternum dull and unpunctured. Elytra with scutellary striae, third interval with three dorsal punctures. Wings usually absent. In marshes, shaded or damp grassland and upland moors; iv-viii. Widespread in Britain and Ireland; very abundant.
Similar species: See *P. minor* (14) and *P. vernalis* (17). Closest to *P. strenuus* (19) but sides of prosternum dull and unpunctured; pronotal margin only slightly sinuate just in front of hind angles, and legs darker, with at least femora dark brown or black.

19. *Pterostichus strenuus* **(Panzer)** Plate 86

Length 6-7.2 mm. Black, legs reddish brown, antennae dark with basal three segments reddish. Pronotal sides distinctly sinuate well in front of the hind angles (Fig. 171); basal punctuation variable, sometimes extensive or restricted to around the single, linear foveae or almost lacking. Prosternum shining, at least partly punctured above the front coxae (Fig. 172). Elytra with scutellary striae, third interval with three dorsal punctures. In almost all habitats except at high altitudes, especially grasslands; iv-viii. Widespread in Britain and Ireland; very abundant.
Similar species: See *P. minor* (14), *P. vernalis* (17) and *P. diligens* (18).

4. *ABAX* Bonelli

A very distinctive genus of large, wide and flat carabids, with the seventh elytral interval keeled behind the toothed shoulder (Fig. 151) (except in one European species) and without dorsal elytral punctures. There are more than ten European species; only one is resident in Britain and Ireland, although a second, *A. parallelus* (Duftschmid), has been recorded as a casual immigrant in the Isles of Scilly.

1. *Abax parallelepipedus* (Piller & Mitterpacher) Plate 87

Length 17-22 mm. Wide and flat, entirely shining black including appendages. Claw segments with ventral setae. Head with doubled and wrinkled furrows inside eyes. Pronotum widest in front of middle, sides behind this almost straight and converging to sharp, rectangular hind angles; hind margin concave; foveae double, linear and unpunctured. Elytral shoulders strongly toothed; with eight fully developed strong striae, intervals slightly convex; third interval without dorsal punctures; seventh interval raised into a sharp keel anteriorly and apically. Wings absent. In woods and damp, well vegetated moorlands; v-ix. Widespread in Britain, except in north-east Scotland, and in Ireland; abundant.

Tribes SPHODRINI and PLATYNINI

A very large, world-wide group of rather flat, active carabids, formerly classed within the Pterostichini but distinguished from that tribe by the more delicate form, less dilated front tibiae (Fig. 39) and the elytral margin not crossing the epipleuron near apex (Fig. 40). There are two setiferous punctures inside the eyes, and the penultimate labial palpal segment has two or fewer setae. The separation of the Sphodrini and Platynini is based largely on features of the male genitalia, so the two are here considered as one for ease of identification using external features. Genera 1-5 in the following key are in the Sphodrini, the remainder are Platynini. The large genus *Agonum* is now divided into seven genera.

Key to genera of Sphodrini and Platynini

1. Length more than 20 mm .... 4. *Sphodrus* Clairville (p. 122)

- Length less than 18 mm 2

2. Tarsi pubescent above (Fig. 173) ... 5. *Laemostenus* Bonelli (p. 123)

- Tarsi not pubescent above 3

3. Claws toothed internally (Fig. 174) 4

- Claws smooth 5

4. Apical segment of labial palpi swollen (Fig. 175); sides of pronotum almost uniformly rounded, hind angles hardly distinct (Plate 89) .... 2. *Synuchus* Gyllenhal (p. 118)

- Apical segment of labial palpi parallel-sided (Fig. 176); sides of pronotum quite or almost straight in front of hind angles, which are distinct and often sharp (Plates 90, 91) .... 3. *Calathus* Bonelli (p. 118)

5. Pronotum rectangular, foveae sharp and linear and anterior margin produced medially (Plate 88) .... 1. *Platyderus* Stephens (p. 118)

- Pronotum shaped differently, foveae wider or indistinct, anterior margin not or less produced 6

6. Sides of pronotum sinuate in front of relatively sharp hind angles (Plates 95-98) 7

- Sides of pronotum not sinuate, hind angles rounded or indistinct (Plates 99-101) 10

7. Third antennal segment pubescent apically; pronotum green or blue (Plate 97) .... 9. *Anchomenus* Bonelli (p. 124)

- Third antennal segment not pubescent; pronotum brown or black 8

8. Length more than 10 mm; body and legs black or nearly so (Plate 98) .... 10. *Platynus* Bonelli (p. 125)

- Length less than 9 mm; not entirely dark 9

9. Length less than 7 mm; head black, darker than pronotum and elytra; apex of antennae darker than base (Plate 95) .... 7. *Oxypselaphus* Chaudoir (p. 124)

- Length at least 7 mm; body uniformly dark to mid brown, antennae uniformly very pale (Plate 96) .... 8. *Paranchus* Lindroth (p. 124)

10. Third elytral interval with three to five deeply foveate punctures. .... 12. *Sericoda* Kirby (p. 125)

- Third elytral interval with non-foveate punctures 11

11. Mentum without median tooth; body uniformly reddish bronze, legs pale yellow; pronotum almost as wide as elytra, sides strongly rounded and hind angles not evident (Plate 94) .... 6. *Olisthopus* Dejean (p. 123)

- Mentum with median tooth (Fig. 177); without that combination of colour and pronotal shape 12

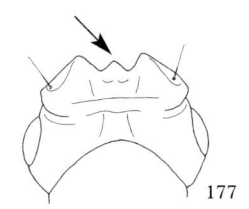

177

12. Head with a transverse impression in front of neck; pronotum not wider than long (Plate 99) and third antennal segment not pubescent .... 11. *Batenus* Motschulsky (p. 125)

- Without impression in front of neck; pronotum usually transverse, if elongate the third antennal segment is pubescent ....13. *Agonum* Bonelli (p. 126)

1. PLATYDERUS Stephens

A genus of about 30 European species, mostly in the Mediterranean region and the Iberian peninsula. There is a single British and Irish species, characterised mainly by the form of the pronotum.

1. *Platyderus depressus* (Audinet-Serville) Plate 88

Length 6-8.5 mm. Body very flat, uniformly reddish brown, head sometimes slightly darker. Appendages pale red-brown; claws long and not toothed. Pronotum almost as long as wide, middle of front margin protruding forwards and angled; foveae sharp and linear, hind angles rounded but distinct. Transverse basal border of elytra usually extending inwards to meet the scutellum; striae not punctured, third interval with three dorsal punctures. Wings usually present. In fields, gardens and waste ground in open situations on dry soils; vi-ix. Widespread in England, more local in the north and west, with single records from Wales and Ireland; scarce.

2. SYNUCHUS Gyllenhal

This genus of over 50 mainly Oriental species resembles *Calathus* but with a more rounded pronotum and swollen apical labial palpal segment (Fig. 175). There is a single species in western Europe.

1. *Synuchus vivalis* (Illiger) Plate 89

Length 6-8.5 mm. Uniformly mid to dark brown, non metallic, elytra with fine transverse microsculpture. Appendages light reddish brown; last segment of labial palpi swollen, hatchet-shaped (Fig. 175); tarsal claws toothed. Pronotum convex, sides almost evenly rounded, hind angles rounded, very obtuse, marked by a setiferous puncture. Elytral striae unpunctured, intervals slightly convex, third interval with two dorsal punctures. Wings usually present. In gardens, grassland, open woodlands and arable land; vii-x. Widespread in Britain, local in Ireland; sometimes abundant.

3. CALATHUS Bonelli

This world-wide genus of over 100 species is characterised by the toothed claws (Fig. 174) and non-swollen apical segment of the labial palpi (Fig. 176). Generally rather flat beetles, many found in dry habitats. There are eight species in Britain, seven in Ireland. One species, *C. cinctus* Motschulsky, has only been recognised relatively recently.

Key to species of *Calathus*

1. Sides of pronotum contracted to very rounded hind angles, so that pronotal base is much narrower than elytral width at shoulders (Plate 90) .... 1. *rotundicollis* Dejean

- Sides of pronotum not or less contracted, hind angles less rounded, pronotal base almost as wide as elytral width at shoulders 2

2. Third and fifth elytral intervals both with dorsal punctures (Fig. 178, Plate 91); pronotal foveae strongly punctured .... 5. *fuscipes* (Goeze)

- Only third interval with dorsal punctures; base of pronotum at most very finely punctured 3

3. Pronotal hind angles sharp (Figs 179, 181) 4

- Pronotal hind angles rounded at extreme tip (Figs 183, 185) 5

4. Pronotum widest behind middle, hardly narrowed towards base, which is as wide as elytra at shoulders (Fig. 179); aedeagus without apical disc (side view Fig. 180) .... 2. *ambiguus* (Paykull)

- Pronotum widest at or in front of middle, slightly narrowed to base, which is not quite as wide as elytra at shoulders (Fig. 181); aedeagus with apical disc (side view Fig. 182) .... 4. *erratus* (Sahlberg)

5. Pronotum mostly dark brown or black, not paler than head and elytra; appendages pale 6

Pronotum paler than head and elytra, or if dark, appendages also darkened 7

6. Dark brown to black with pronotal margins paler; pronotal side border widened basally (Fig. 183); tip of right paramere of aedeagus strongly hooked (Fig. 184) .... 7. *micropterus* (Duftschmid)

- Pale to mid brown; pronotal side border not clearly widened basally (Fig. 185); tip of right paramere of aedeagus slender, constricted before apex which is completely rounded or with a small tooth (Fig. 186) .... 8. *mollis* (Marsham)

7. Head and elytra black; elytra more convex, so that eighth interval appears narrow when viewed from above, shoulders with a minute protruding tooth (Fig. 187); tip of right paramere of aedeagus toothed (Fig. 188) .... 6. *melanocephalus* (Linnaeus)

- Head and elytra brown; elytra less convex so that eighth interval appears wider from above, shoulders not toothed (Fig. 189); tip of right paramere of aedeagus broad, not constricted before apex, which is not or hardly toothed (Fig. 190) .... 3. *cinctus* Motschulsky

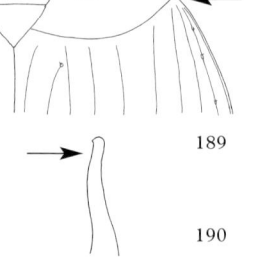

Subgenus *Amphigynus* Haliday

1. *Calathus rotundicollis* Dejean Plate 90

Length 8.5-10.5 mm. Body and appendages almost uniformly mid to dark brown, with pronotal and elytral margins, antennae and tarsi paler. Pronotum strongly contracted to completely rounded hind angles, base narrower than elytral shoulder width. Elytra strongly microsculptured, striae unpunctured; third interval with four or five dorsal punctures usually touching third stria. Wings present or absent. In woodlands, also sometimes gardens and coastal dunes; v-x. Widespread throughout Britain, except in the extreme north of Scotland, and Ireland; often abundant.

Subgenus *Calathus sensu stricto*

2. *Calathus ambiguus* (Paykull)

Length 8.5-11.5 mm. Black or dark brown, pronotum slightly paler than elytra, especially marginally. Appendages pale brown or yellow. Pronotum wide, not narrowed in basal half, basally as wide as elytral shoulders (Fig. 179). Elytral intervals very flat, third interval with two dorsal punctures. Aedeagal tip abruptly curved, without apical disc (Fig. 180). Wings present. In open, dry sandy sites, both inland and on the coast; vi-viii. Local in England and Wales, mainly eastern; scarce.
Similar species: Differs from *C. erratus* (4) in having a wider pronotum which is not narrowed basally, and by the lack of apical disc on the aedeagus. On average smaller than *C. fuscipes* (5) which has a punctured pronotum and dorsal punctures on the fifth elytral interval.

3. *Calathus cinctus* Motschulsky

Length 6-8.5 mm. Head and elytra mid to dark reddish brown with paler margins, pronotum bright reddish. Appendages pale red or yellow-brown. Pronotal sides slightly rounded, hind angles not sharp. Elytra rather flat, shoulders not toothed (Fig. 189), sides curved and only slightly sloping downwards, eighth interval appearing as wide as elytral border when viewed from above. Right aedeagal paramere broad, its apex untoothed or nearly so (Fig. 190). Wings present or absent. On coastal dunes and dry lowland heaths; vii-x. Widespread but local in England; very local around the coasts of Wales, Scotland and Ireland; sometimes abundant.
Similar species: Flatter than *C. melanocephalus* (6) and less brightly coloured; the pronotum is always paler than the head and elytra, which distinguishes it from *C. mollis* (8).

4. *Calathus erratus* (Sahlberg)

Length 8.5-11.5 mm. Black, margins of pronotum paler. Appendages pale brown. Pronotum widest at middle, slightly narrowed basally; base clearly narrower than elytral shoulder width; basal foveae distinct and finely punctured (Fig. 181). Elytral intervals often slightly convex anteriorly, third interval with two dorsal punctures. Aedeagal tip with apical disc (Fig. 182). Wings present or absent. In open, dry sandy sites and coastal dunes; vii-x. Widespread in Britain; local and exclusively coastal in Ireland; abundant.
Similar species: See *C. ambiguus* (2). Smaller than *C. fuscipes* (5) and with dorsal punctures only on the third elytral interval.

5. *Calathus fuscipes* (Goeze) Plate 91

Length 10-14 mm. Black, head and pronotum shining, elytra duller with strong microsculpture. Antennae brown sometimes with middle segments darkened; legs usually light brown with tarsi darkened, sometimes all black or dark brown. Pronotum only slightly transverse, sides slightly rounded and contracted in front of hind angles; base slightly narrower than elytral shoulder width; basal foveae rough with some large punctures. Elytral intervals convex, third and fifth intervals with dorsal punctures (Fig. 178). Wings absent. In open grasslands, arable fields and gardens; vi-x. Widespread in Britain and Ireland; very abundant.
Similar species: See *C. ambiguus* (2) and *C. erratus* (4).

6. *Calathus melanocephalus* (**Linnaeus**)

Length 6-8.5 mm. Head and elytra slightly metallic, very dark brown or black, pronotum typically bright reddish. Appendages pale red or yellow-brown. Pronotal side borders not widened in front of the slightly rounded hind angles. Elytra convex, shoulders with a minute protruding tooth (Fig. 187), sides straight and distinctly sloping downwards laterally, eighth interval appearing narrower than elytral border when viewed from above. Tip of right aedeagal paramere clearly toothed (Fig. 188). Wings usually absent. The var. *nubigena* Haliday is all black, with all or most of the legs also darkened. In grasslands, heaths, gardens and arable land; vi-x. The typical form occurs at all altitudes, var. *nubigena* only in upland areas. Widespread in Britain and Ireland; very abundant.
Similar species: See *C. cinctus* (3). The dark form can be distinguished from *C. mollis* (8) and *C. micropterus* (7) by having dark appendages.

7. *Calathus micropterus* (**Duftschmid**)

Length 6.5-8.5 mm. Head and elytra black, pronotum with paler margins; appendages pale yellow-brown. Pronotum hardly transverse, side borders widened in front of the rounded hind angles (Fig. 183). Elytra moderately convex, and with rounded sides. Apex of right aedeagal paramere strongly hooked (Fig. 184). Wings absent. In upland forests and heather moors; v-x. Widespread in the upland regions of Britain, especially in the east; local in Ireland; sometimes abundant.
Similar species: See *C. melanocephalus* (6). Distinguished from both that species and *C. mollis* (8) by the widened pronotal side border in front of the hind angles. The pronotum is also less transverse than in *C. mollis*.

8. *Calathus mollis* (**Marsham**)

Length 6.5-9 mm. More or less uniformly pale to mid reddish brown, sometimes paler marginally. Appendages pale yellow-brown. Pronotum transverse, side borders not widened behind (Fig. 185). Elytra rather flat, with rounded sides. Right aedeagal paramere slender, constricted before apex which is rounded and almost completely untoothed (Fig. 186). Wings present. In coastal dunes and occasionally in sandpits inland. Widely distributed around the coasts of Britain and Ireland; very abundant.
Similar species: See *C. cinctus* (3), *C. melanocephalus* (6) and *C. micropterus* (7).

4. *SPHODRUS* Clairville

A genus of two very large species, characterised by the very long third antennal segment. There is a single British and Irish species.

1. *Sphodrus leucophthalmus* (**Linnaeus**) Plate 92

Length 21-28 mm. Body and appendages uniformly dull black or dark brown, strongly microsculptured. Third antennal segment much longer than first two segments combined. Legs exceptionally long. Pronotum strongly narrowed and sinuate in front of sharp hind angles, base finely punctured. Elytral striae very finely punctured, faint towards sides; third interval without dorsal punctures. Wings present. In cellars, stables and old buildings, probably introduced. Extremely local in England and south Wales; presumed extinct in Ireland where it has not occurred since the 19th century; extremely scarce.
Similar species: Resembles a large *Pterostichus niger* in size and black colour but more convex with slenderer front tibiae, pronotum more cordate and striae finer.

5. *LAEMOSTENUS* Bonelli

A Palaearctic genus of large, rather flat beetles with pubescent tarsi (Fig. 173). Superficially resembling some *Pterostichus* but with the epipleuron not crossed by the elytral margin (Fig. 40), and elytra without dorsal punctures. There are over 150 species, sometimes divided into separate genera. Only two species occur in Britain and Ireland, and one of these is an introduction.

Key to species of *Laemostenus*

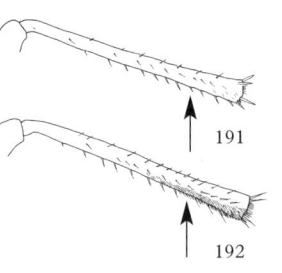

1. Inner side of hind tibiae pubescent only near apex, or not at all (Fig. 191) ....1. *complanatus* (Dejean)

- Inner side of hind tibiae densely pubescent throughout apical half (Fig. 192) .... 2. *terricola* (Herbst)

Subgenus *Laemostenus sensu stricto*

1. *Laemostenus complanatus* (Dejean)

Length 12.5-16 mm. Black, elytra sometimes slightly bluish; appendages black or dark brown. Hind tibiae straight in both sexes, almost without pubescence on inner face (Fig. 191). Eyes bulging, longitudinal diameter as long as temples. Pronotum quadrate or slightly transverse, sides sinuate just in front of the sharp hind angles; basal foveae deep, usually slightly roughened. Wings present. An introduced species, found in coastal litter often near ports; v-ix. Local but widespread in England and south Wales, usually coastal; one Scottish record near Edinburgh; very local in Ireland; very scarce.
Similar species: Distinguished from *L. terricola* (2) by the lack of most pubescence on the inner face of the hind tibiae; also the eyes are more prominent and the male mid tibiae are straight.

Subgenus *Pristonychus* Dejean

2. *Laemostenus terricola* (Herbst) Plate 93

Length 13-17 mm. Black, elytra bluish, appendages black or nearly so. Hind tibiae with apical half of inner face densely pubescent (Fig. 192); male mid tibiae strongly curved inwards apically. Eyes not protruding, their longitudinal diameter shorter than the temples. Pronotum quadrate to slightly elongate, rather flat, sides sinuate some way in front of the hind angles; basal foveae shallow, smooth. Wings absent. In woods, gardens, buildings and associated with mammalian burrows; vi-x. Widespread but local in England and Ireland; mainly coastal in Wales and Scotland; usually scarce.
Similar species: See *L. complanatus* (1).

6. *OLISTHOPUS* Dejean

A Holarctic genus, with five European species but only one in Britain and Ireland. Superficially similar to *Synuchus* but distinguished by the smooth claws and metallic bronze colour.

1. *Olisthopus rotundatus* (Paykull) Plate 94

Length 6.5-8 mm. Body reddish brown with a distinct bronze reflection. Appendages yellow with tarsi and antennal segments 4-11 somewhat darkened. Mentum without a prominent median tooth. Pronotum transverse with sides strongly rounded and hind angles not distinct; indistinct basal foveae sometimes finely punctured. Elytral striae unpunctured, third interval with three dorsal punctures. Wings present. On dry moors and heaths, usually where there is *Calluna* but also on coastal dunes; v-x. Widespread in Britain and Ireland; sometimes abundant.

7. OXYPSELAPHUS Chaudoir

This and the following three genera were all considered part of *Agonum* but are distinguished by their relatively sharp pronotal hind angles. There is a single species in this genus, characterised by the very narrow and elongate pronotum, smooth third antennal segment and non-furrowed tarsi.

1. *Oxypselaphus obscurus* (Herbst) Plate 95

Length 5-6.5 mm. Head black, pronotum and elytra mid to dark brown. Appendages pale, except antennae darkened from fourth segment to apex. Pronotum no wider than head, elongate, sides very narrowly bordered and strongly sinuate, basal third strongly punctured. Elytral striae strongly punctured, third interval usually with three very fine dorsal punctures. Wings absent. In marshes and damp shaded habitats including woodland; iv-viii. Widespread in England, local in Wales and Ireland; sometimes abundant.

8. PARANCHUS Lindroth

Separated from *Anchomenus* by the non-pubescent third antennal segment. The dorsal grooves on the hind tarsi distinguish the genus from *Oxypselaphus* and *Platynus*. There is a single widely distributed western European species, as well as two others in Madeira.

1. *Paranchus albipes* (Fabricius) Plate 96

Length 6.5-8.8 mm. Uniformly mid to dark brown. Appendages pale yellow; tarsi with median dorsal groove. Pronotum slightly transverse, sides contracted and sinuate just in front of the slightly protruding, sharp and setate hind angles; foveae shallow, indistinct, much of base strongly punctured. Elytral striae deep and unpunctured, intervals concave, third interval with two dorsal punctures. Wings present. At the margins of all types of fresh water; iv-x. Widespread in Britain and Ireland; very abundant.

9. ANCHOMENUS Bonelli

Distinguished by the pubescent third antennal segment. *Anchomenus* is Holarctic, with about 12 species but only one in western Europe.

1. *Anchomenus dorsalis* (Pontoppidan) Plate 97

Length 6-8 mm. Head and pronotum metallic green or blue, elytra reddish brown with apical half green or blue except at margins; rarely elytra entirely green. Appendages pale brown, apical segments of antennae darker; third antennal segment mostly pubescent;

tarsi not grooved dorsally. Pronotum rather flat, sides contracted and sinuate in front of the obtuse but almost sharp and unpunctured hind angles; foveae elongate, surface finely punctured between these and side margins. Elytral striae deep and unpunctured, intervals flat, third interval with four fine punctures. Wings present. In arable fields, gardens and on waste ground with dry soils; v-viii. Widespread in Britain, except in northern Scotland, and in Ireland; sometimes very abundant, forming large aggregations.

10. *PLATYNUS* Bonelli

As recognised here, this genus has only three European species but there are hundreds of species in the Americas. The single British and Irish species is characterised by its large size, black colour, sinuate pronotal sides with sharp hind angles, smooth third antennal segment and non-grooved tarsi.

1. *Platynus assimilis* (Paykull) Plate 98

Length 9-12.5 mm. Body entirely shining black. Appendages mid to dark brown, femora and antennal bases black. Pronotum transverse, sides contracted and sinuate in front of sharp, protruding and raised hind angles; side borders wide and strongly raised. Elytra wide, sides sinuate near apices which are produced; striae finely punctured, intervals convex and with transverse microsculpture; third interval with three deep dorsal punctures. Wings present. Usually in woods, especially near water; v-x. Widespread in Britain and Ireland; very abundant.

11. *BATENUS* Motschulsky

A single species, characterised by the constricted neck and pronotal shape: formerly a subgenus of *Platynus* or *Agonum*.

1. *Batenus livens* (Gyllenhal) Plate 99

Length 7.5-10 mm. Black or dark brown, head usually with two faint red spots dorsally between the eyes. Appendages mid to dark brown, antennae paler basally; tarsi not furrowed. Head constricted behind the eyes, with a transverse impression in front of the neck. Pronotum hardly transverse, narrowed basally, side borders narrow in front, widened and raised basally. Elytra rather long and flat, widest behind middle; intervals convex. Wings present. In lowland marshes and fens, sometimes under bark; iii-vi. Very local in southern and eastern England and in southern Ireland; sometimes abundant.
Similar species: Resembles *Agonum (Europhilus) thoreyi* because of the elongate pronotum but is larger, with non-pubescent third antennal segment.

12. *SERICODA* Kirby

Formerly a subgenus of *Agonum* but distinguished by the deep, foveate dorsal elytral punctures. There are seven species world-wide, most associated with burnt forests; only two are European and one British.

1. *Sericoda quadripunctata* (De Geer)

Length 4.5-5.7 mm. Black with a slight bronze metallic sheen, and strong iridescent microsculpture, especially on the elytra. Appendages black, tibiae sometimes slightly paler. Pronotum transverse, anterior margin often concave, front angles protruding forwards, sides rounded and contracted to the very obtuse but just discernible hind angles; strongly raised side borders widening towards base. Elytral surface rather irregular, striae fine and intervals convex; third interval with four strongly foveate punctures. Wings present. On burnt ground, both forests and heaths; iv-vi. Scattered records throughout England and Scotland but last recorded in southern England in 1975; extremely scarce.

Similar species: Smaller than most black *Agonum* species, except some of the subgenus *Europhilus*, from which it is distinguished by the foveate elytral punctures and the non-pubescent third antennal segment.

13. *AGONUM* Bonelli

This enormous group, with hundreds of species world-wide, is sometimes further split into separate genera. The subgenus *Europhilus* in particular has often been given generic status and is keyed separately here for convenience. All *Agonum* have a toothed mentum (Fig. 177) and rounded or very obtuse pronotal hind angles. Most are associated with damp habitats. There are 15 British species, 12 in Ireland (one of which does not occur in Britain); some are distinctive but others can be difficult to identify.

Key to subgenera of *Agonum*

1. Apical half or more of third antennal segment pubescent (Fig. 193) .... 1. *Europhilus* Chaudoir (p. 126)

- Third antennal segment not pubescent .... 2. *Agonum sensu stricto* (p. 129)

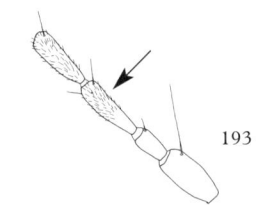

1. Subgenus *Europhilus* Chaudoir

A Holarctic subgenus of more than 25 usually marsh-living species that can be difficult to identify. Five of the six British species also occur in Ireland.

Key to species of *Agonum* (*Europhilus*)

1. Pronotum narrow, almost elongate (Fig. 194); hind tarsi with a median dorsal furrow (Fig. 195) .... 6. *thoreyi* Dejean

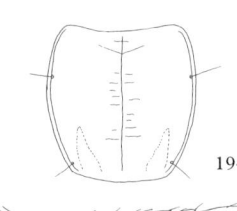

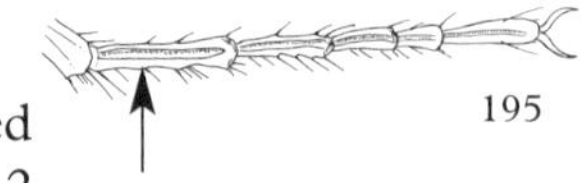

- Pronotum at least 1.2 times as wide as long; hind tarsi furrowed laterally or not at all 2

2. Head and pronotum black, elytra yellowish bronze; pronotum very small in comparison to elytra (Fig. 196) .... 4. *piceum* (Linnaeus)

- Head and pronotum not so much darker than elytra; pronotum larger 3

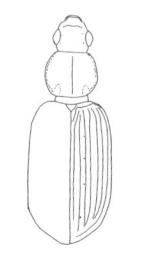

3. Side borders of pronotum uniformly narrow, front margin of elytra meeting shoulder at a sharp angle (Fig. 197); upper surface not or hardly metallic 4

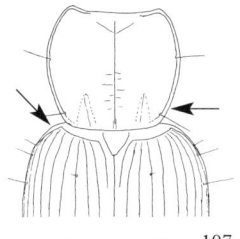

- Side borders of pronotum widening in front of hind angles, front margin of elytra curving to meet shoulder (Fig. 198); upper surface with a distinct metallic reflection 5

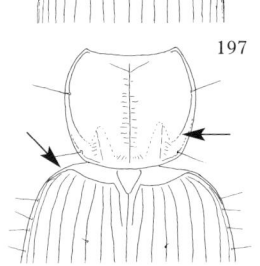

4. Upper surface and appendages entirely black; wings present .... 2. *gracile* Sturm

- At least elytra and/or tibiae somewhat tinged with red or brown (Plate 100); wings usually absent .... 1. *fuliginosum* (Panzer)

5. Dorsal surface with a bronze or greenish reflection; pronotal sides rounded, moderately contracted behind (Fig. 198) .... 3. *micans* Nicolai

- Dorsal surface with a bluish reflection; pronotal sides strongly contracted and almost straight in basal half (Fig. 199) .... 5. *scitulum* Dejean

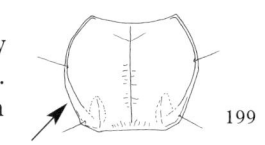

1. *Agonum fuliginosum* (Panzer) Plate 100

Length 5.5-7 mm. Non-metallic; head black, pronotum black, sometimes paler marginally, elytra black or dark reddish brown. Antennae black or with first segment slightly paler; legs mid to dark brown, femora often almost black. Pronotum about 1.2 times as wide as long, side borders uniformly narrow (Fig. 197). Elytra short and convex, together 1.5 times as long as wide, less than three times as long as pronotum, sides rounded; anterior margin meeting shoulder at a sharp angle (Fig. 197). Wings usually absent, rarely present. In marshes, damp grasslands and moorland; iv-ix. Widespread in Britain and Ireland; very abundant.

Similar species: *A. gracile* (2) has a similarly narrow-bordered pronotum but is entirely black; all other species of the subgenus *Europhilus* have relatively smaller pronota and all except *A. scitulum* (5) have longer, less rounded elytra. *A. thoreyi* (6) can be similarly coloured, but has a more elongate pronotum and furrowed hind tarsi. See also *A. nigrum* (13).

2. *Agonum gracile* Sturm

Length 5.8-7.1 mm. Body shining black or slightly metallic, appendages black. Pronotum 1.2 times as wide as long, side borders narrow throughout. Elytra long, together 1.6 times as long as wide and more than three times as long as pronotal length, almost parallel-sided. Wings present. In marshes, bogs and on upland acid grasslands and moors; v-viii. Widespread in Britain and Ireland; abundant.

Similar species: See *A. fuliginosum* (1). *A. micans* (3) and *A. scitulum* (5) are more metallic, and have the pronotal side borders widened in front of the hind angles.

3. *Agonum micans* Nicolai

Length 6.2-7.3 mm. Head and pronotum metallic black, elytra dark reddish or brown, all with a bronze-black or greenish metallic reflection. Antennae black with first segment brown; legs brown. Eyes three times as long as temples. Pronotum 1.25 times as wide as long, with sides rounded, side borders widened but hardly raised in front of hind angles (Fig. 198), basal foveae shallow. Elytra together more than 1.5 times as long as wide, 3.3 times as long as pronotum, sides straight and slightly widening to apical third; anterior margin meeting shoulder without a sharp angle (Fig. 198). Wings present. In damp, shaded situations near water, especially carr woodland, where it overwinters under bark; iv-vii. Local in England and Wales, extremely local in Scotland and Ireland; sometimes abundant.
Similar species: See *A. fuliginosum* (1) and *A. gracile* (2). *A. scitulum* (5) is bluish rather than bronze, and has smaller eyes; its pronotum is more constricted basally and the hind angles are more raised with deeper foveae.

4. *Agonum piceum* (Linnaeus)

Length 5.5-7.1 mm. Head and pronotum black with a slight metallic reflection; elytra bronze-yellow or brown, often showing the folded wings beneath. Antennae black with first segment hardly paler; legs brown. Pronotum small, often rather rectangular, about 1.25 times as wide as long; side borders somewhat widened in front of hind angles. Elytra long and parallel-sided, almost 3.5 times as long as pronotum and together about 1.7 times as long as wide (Fig. 196). Wings present. In lowland fens, bogs and marshes on heavy soils; iv-viii. Local throughout Britain and Ireland; sometimes abundant.
Similar species: Only *A. thoreyi* (6) has as great a colour contrast between fore-parts and elytra but the pronotum of that species is larger relative to the elytra and more elongate.

5. *Agonum scitulum* Dejean

Length 5.6-7.1 mm. Shining black, elytra with a slight bluish reflection; appendages black except tibiae reddish brown. Eyes twice as long as the temples. Pronotum 1.25 times as wide as long, widest in front of middle and rather strongly narrowed towards the obtuse hind angles; side borders widened and strongly raised in front of these (Fig. 199), with foveae distinct. Elytra rather short, together 1.5 times as long as wide, with sides strongly rounded. Wings probably present. In shaded marshes and carrs; v-vii. Extremely local in England and Wales; very scarce.
Similar species: See *A. fuliginosum* (1), *A. gracile* (2) and *A. micans* (3).

6. *Agonum thoreyi* Dejean

Length 6-8 mm. Head black, pronotum brown or black with paler margins, elytra yellowish brown or dark brown to almost black. Antennae dark with basal segments paler; legs light to dark brown, tarsi with a median dorsal furrow (Fig. 195). Pronotum quadrate, sides evenly rounded, or straight in posterior half; side borders uniformly narrow, foveae virtually lacking (Fig. 194). Elytra together more than 1.6 times as long as wide, sides slightly rounded, widest behind middle. Wings present. The type form is pale, the dark form is *puellum* Dejean. In marshes, fens and reed beds; iii-viii. Widespread in England, Wales, southern Scotland and Ireland; often abundant.
Similar species: See *A. fuliginosum* (1) and *A. piceum* (4).

Subgenus *Agonum sensu stricto*

This very diverse subgenus contains 10 British or Irish species all distinguished from subgenus *Europhilus* by their non-pubescent third antennal segment. Nine of them are found in Britain and seven in Ireland.

Key to species of *Agonum sensu stricto*

1. Elytra metallic golden-green with pale yellow margins .... 11. *marginatum* (Linnaeus)

- Differently coloured 2

2. Third elytral interval with four to six punctures; side borders and foveae of pronotum roughly sculptured or punctured (Fig. 200) 3

- Third elytral interval with three punctures; side borders and foveae of pronotum smooth 5

3. Head and pronotum bright metallic green, in contrast to the reddish-coppery elytra .... 14. *sexpunctatum* (Linnaeus)

- Almost uniformly coloured, either green, blue, purple, coppery, bronze or black, with little contrast between pronotum and elytra 4

4. First antennal segment and tibiae mid to pale brown; body uniformly bronze coloured .... 9. *gracilipes* (Duftschmid)

- Appendages all black; body variably coloured .... 8. *ericeti* (Panzer)

5. First antennal segment (and often tibiae, at least basally) mid to dark brown 6

- First antennal segment and tibiae black 8

6. Head and pronotum usually with a green reflection, elytra dark coppery (Plate 101); elytral striae extremely fine and intervals completely flat .... 12. *muelleri* (Herbst)

- Head and pronotum black, or with a bronze reflection; elytral striae deeper, intervals clearly convex apically 7

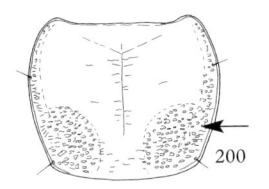

7. Pronotum about 1.2 times as wide as long, with very narrow side borders, especially at front (Fig. 201) .... 13. *nigrum* Dejean

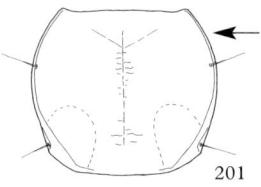

- Pronotum about 1.4 times as wide as long, with wider side borders (Fig. 202) .... 15. *versutum* Sturm

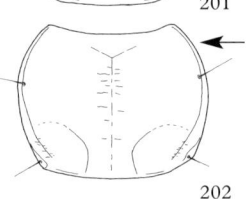

8. Hind angles of pronotum present as a slight protruding angle, elytra longer in proportion to pronotum with basal margin more curved near scutellum (Fig. 203) .... 10. *lugens* (Duftschmid)

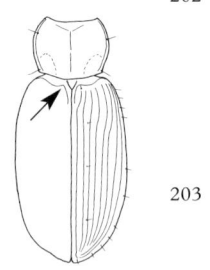

- Hind angles of pronotum completely rounded, elytra shorter with basal margin less curved (Fig. 204) 9

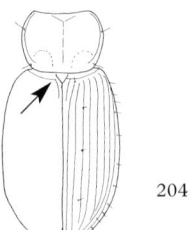

9. Upper surface black, non metallic; apex of aedeagus curved downwards in side view (Fig. 205) ...... 7. *emarginatum* (Gyllenhal)

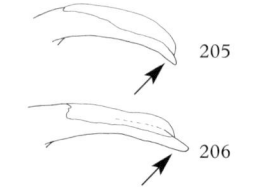

- Upper surface with metallic reflection; apex of aedeagus straight in side view (Fig. 206) .... 16. *viduum* (Panzer)

7. *Agonum emarginatum* (Gyllenhal)

Length 7.5-9 mm. Shining black, non-metallic. Appendages black, base of antennae, tibiae or tarsi sometimes brownish. Pronotum transverse, sides almost regularly rounded, side borders well defined but usually not deepened adjacent to the raised discal area; foveae shallow and almost unpunctured (Fig. 204). Elytra distinctly and moderately transversely microsculptured, shoulders sometimes slightly raised; striae deep and intervals convex apically; third interval with three punctures. Aedeagus with apex curved downwards (Fig. 205). Wings present. In marshes, and near fresh water; iv-viii. Widespread in Britain and Ireland except in Scotland, where it is very local; abundant.
Similar species: Only separable from *A. viduum* (16) by the non metallic black upper surface, and curved apex of the aedeagus. The smooth pronotum distinguishes it from black examples of *A. ericeti* (8). See also *A. lugens* (10).

8. *Agonum ericeti* (Panzer)

Length 6.5-8 mm. Typically metallic coppery or brassy with greenish reflections especially on the head and pronotum but can be bluish, purple or even black. Appendages black. Pronotum transverse with rounded sides and wide, roughened side borders and basal foveae (Fig. 200). Elytral striae fine, intervals flat; third interval with four to seven punctures. Wings absent. In bogs and on wet heaths, usually with *Sphagnum* mosses; v-vii. Local in Hampshire and Dorset heaths, Wales, northern England and Scotland; usually scarce.

Similar species: *A. sexpunctatum* (14) has a similarly roughened pronotal base and sides but that species is larger, with more colour contrast between the fore parts and the elytra. *A. muelleri* (12) also has flat elytral intervals, but the pronotum is not roughened. Bronze specimens of *A. ericeti* are distinguished from *A. gracilipes* (9) by their black appendages. See also *A. emarginatum* (7).

9. *Agonum gracilipes* (Duftschmid)

Length 7-8.5 mm. Uniformly metallic bronze, female dull from dense microsculpture, male more shining. Appendages dark brown, first antennal segment and tibiae paler brown. Pronotal foveae and side borders often roughened and with a trace of punctures; pronotal sides contracted to a slightly protruding tooth at obtuse hind angles. Anterior margin of elytra strongly arched; striae fine, intervals flat; third interval with four to seven punctures. Wings present. Only known from occasional single immigrant specimens, mainly found near the coast of eastern England but also known from Wales and western Scotland.

Similar species: See *A. ericeti* (8).

10. *Agonum lugens* (Duftschmid)

Length 8-10 mm. Non-metallic black, strongly microsculptured. Appendages black or very dark brown, all segments of mid and hind tarsi furrowed laterally. Pronotum transverse, rather flat, with hind angles very obtuse but distinct and sometimes with a small tooth. Elytra long (Fig. 203), anterior margin strongly curved forwards to meet scutellum and at shoulders; surface sometimes strongly raised inside shoulders; intervals convex, third interval with three punctures. Wings present. On silty soil near fresh water; v-viii. Only known from Clare, Ireland; sometimes abundant.

Similar species: The non-metallic black surface and appendages resemble *A. emarginatum* (7) but *A. lugens* has longer elytra relative to the pronotal length, more distinct pronotal hind angles, more curved elytral front margins and all segments of the hind tarsi furrowed.

11. *Agonum marginatum* (Linnaeus)

Length 8.8-10.4 mm. Bright coppery green, margins of elytra pale yellow or cream. Appendages dark metallic brown, with tibiae and first antennal segment paler. Pronotum transverse with sides evenly rounded. Elytra strongly microsculptured; striae very fine and not punctured, intervals convex, third interval with three punctures. Wings present. In wet, vegetated lowland habitats usually near water, also on the coast; v-viii. Widespread in the south of England, Wales and Ireland, local in northern England and Scotland; abundant.

12. *Agonum muelleri* (Herbst) Plate 101

Length 7-9 mm. Head and pronotum dark coppery green, elytra dark coppery or brassy; rarely entirely black. Appendages dark brown to black, with tibiae (except apices) and first antennal segment paler. Pronotum only slightly transverse, sides evenly rounded, not roughened. Elytral striae very fine, especially apically, intervals flat; third interval with three punctures. Wings present. In damp grasslands, fields, gardens, open woodland and dune slacks; iv-viii. Widespread in Britain and Ireland; abundant.
Similar species: The colours are much duller than in *A. sexpunctatum* (14) which has a roughened pronotal base and more dorsal elytral punctures. Rare black examples can be recognised from *A. emarginatum* (7) by the pale first antennal segment, and from *A. nigrum* (13) and *A. versutum* (15) by the fine elytral striae and flat intervals, especially apically. See also *A. ericeti* (8).

13. *Agonum nigrum* Dejean

Length 7-9 mm. Black or dark brown, elytral suture sometimes paler. Appendages dark brown to black, bases of antennae usually slightly paler. Pronotum hardly transverse, sides contracted basally, side borders extremely narrow (Fig. 201). Elytra widest behind middle, intervals convex, microsculpture consisting of slightly transverse meshes; third interval with three or (rarely) four punctures. Wings present. In vegetated marshes, usually coastal; v-viii. Very local in England, Wales, south-west Scotland and Ireland; very scarce.
Similar species: The slender pronotum with narrow side borders resembles *A. (Europhilus) fuliginosum* (1) but that species has a pubescent third antennal segment. *A. versutum* (15) also has a pale antennal base but has a wider pronotum and transverse microsculpture. See also *A. muelleri* (12).

14. *Agonum sexpunctatum* (Linnaeus)

Length 7-9.5 mm. Head and pronotum bright shining green, elytra reddish coppery with metallic green margins. Appendages metallic black. Pronotum very transverse with wide side borders, which are strongly roughened. Elytra strongly microsculptured, striae strong and finely punctured, intervals flat; third interval with four to eight punctures. Wings present. On damp open ground near water on sandy soils in heaths and open woodland; v-vii. Extremely local in southern England but with old records from northern England; very scarce.
Similar species: See *A. ericeti* (8) and *A. muelleri* (12).

15. *Agonum versutum* Sturm

Length 7-8.5 mm. Shining black, sometimes with a slight metallic reflection. Appendages black, tibiae and first antennal segment dark brown. Pronotum very transverse, sides evenly rounded, side borders wide (Fig. 202). Elytra parallel-sided, striae usually punctured, intervals flat or convex, microsculpture very transverse, hardly meshed; third interval with three punctures. Wings present. On vegetated damp ground near fresh water; v-vii. Extremely local in southern England (mainly around the Severn estuary) and southern Ireland (last recorded in 1920); very scarce.
Similar species: See *A. muelleri* (12) and *A. nigrum* (13). Smaller than *A. emarginatum* (7) which does not have the pale first antennal segment.

16. *Agonum viduum* (Panzer)

Length 7.5-9 mm. Black with a distinct metallic greenish or brassy reflection, especially on elytra. Appendages black. Pronotal side borders usually slightly depressed between margin and raised discal area. Elytral shoulders often raised, microsculpture very transverse. Apex of aedeagus straight (Fig. 206). Wings present. In marshes and on well vegetated muddy or silty ground near water; v-viii. Widespread in England (especially the west), Wales and Ireland; very local in Scotland; usually scarce.
Similar species: See *A. emarginatum* (7).

Tribe ZABRINI

A tribe of more than 600 mostly Holarctic, at least partly phytophagous carabids. They are usually rather parallel-sided, sometimes stout species, with a smooth third antennal segment. The elytral margin crosses the epipleuron, as in the Pterostichini but there are no dorsal elytral punctures on the third elytral interval. There are either one or two setiferous punctures inside each eye.

Key to genera of Zabrini

1. With a single setiferous puncture inside the eye (as in Fig. 35) .... 1. *Zabrus* Clairville (p. 133)

- With two setiferous punctures inside each eye (Fig. 34) 2

2. Length usually less than 10 mm; pronotum not, or hardly contracted and sinuate laterally (Plates 103-108); prosternal process with a narrow border (Fig. 207) .... 2. *Amara* Bonelli (p. 134)

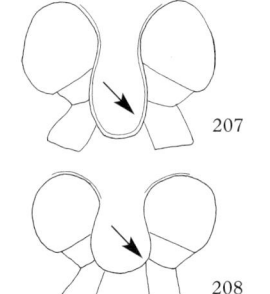

- Length usually more than 10 mm; pronotum strongly contracted and usually sinuate in front of hind angles (Plate 109); prosternal process not bordered (Fig. 208) .... 3. *Curtonotus* Stephens (p. 146)

1. *ZABRUS* Clairville

Although this genus has about 40 European species, most are southern and only one occurs in Britain but not Ireland. It is an occasional pest of cereal crops in the south of England but a major pest in Central Europe.

1. *Zabrus tenebrioides* (Goeze) Plate 102

Length 14-16 mm. A very stout, almost cylindrical species. Dorsal surface black; pronotum with coarse transverse wrinkles, elytra with dense microsculpture. Appendages dark brown with black femora and antennal bases. Front tibiae with a small second apical spur behind the main spur. Pronotal sides strongly bordered, contracted in front, parallel-sided behind; base strongly punctured, foveae shallow. Elytral striae strongly punctured, intervals convex. Wings present. In cereal fields and open, dry grasslands; viii-x. Very local in southern England and south Wales; occasionally abundant.

2. *AMARA* Bonelli

A very large Holarctic genus of about 150 European species, characterised by their rather oval and parallel-sided form; less convex than *Zabrus*, with two setiferous punctures inside the eyes. There are 27 British species (18 in Ireland), excluding three now placed in the separate genus *Curtonotus*. *A. (Celia) cursitans* (Zimmermann) is known only from two specimens near London in the 1950s; it is included in the key only. Females are often duller than the males and may also differ in body shape. Some species have a setiferous puncture at or near the base of the short scutellary stria; this is here referred to as the 'scutellary pore'.

Key to subgenera of *Amara*

1. Apical spur of front tibiae divided into three teeth (Fig. 209) .... 1. *Zezea* Csiki (p. 135)

- Apical spur of front tibiae single (Fig. 210) 2

2. Antennae with one to three pale basal segments, remainder more or less abruptly and uniformly darkened, usually black .... 2. *Amara sensu stricto* (p. 136)

- Antennae uniformly pale brown, or more gradually darkened from segments 3-5 to apex 3

3. Prosternal process without apical setae (as in Fig. 207) 4

- Prosternal process with two to six apical setae (Fig. 211) 5

4. Outer pronotal fovea widened behind and with a distinct oblique ridge externally; hind angles of pronotum protruding or slightly toothed (Fig. 212) .... 5. *Bradytus* Stephens (p. 145)

- Outer pronotal foveae not widened, less raised externally; hind angles of pronotum not or barely toothed (Fig. 213) .... 3. *Celia* Zimmermann (p. 143)

5. Eyes more protruding (Fig. 214); elytral striae deepened apically, intervals convex .... 6. *Percosia* Zimmermann (p. 146)

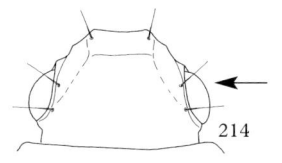

- Eyes rather flat (Fig. 215); elytral striae not deepened apically, intervals flat .... 4. *Paracelia* Bedel (p. 144)

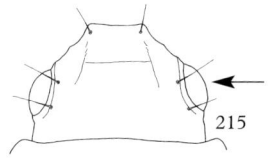

1. Subgenus *Zezea* Csiki

This Holarctic subgenus, with about 20 species, is characterised by the three-toothed spur on the front tibiae (Fig. 209); there is also a scutellary pore as found in some *Amara sensu stricto* and *A. (Celia) praetermissa*. There are only two British species, one of which occurs in Ireland.

Key to species of *Amara* (*Zezea*)

1. Length less than 8 mm .... 1. *plebeja* (Gyllenhal)

- Length at least 8 mm .... 2. *strenua* Zimmermann

1. *Amara plebeja* (Gyllenhal) Plate 103

Length 6-7.8 mm. Black with distinct metallic shine, pronotum usually slightly greenish, elytra more coppery. Antennae black with three basal segments clear red; femora and most of tarsi black, tibiae red or light brown. Pronotal front angles protruding forwards strongly; sides rounded, base punctured, inner foveae sharp, outer foveae deepened inside the backwardly protruding hind angles. Elytral striae rather fine, hardly punctured, intervals convex posteriorly. Wings present. In damp grasslands and other well vegetated, moist habitats including arable fields on heavy soils; iv-viii. Widespread in Britain and Ireland; very abundant.
Similar species: Smaller than *A. strenua* (2), the only other species with a three-pronged tibial spur, and with pronotum more punctured and depressed near hind angles. The combination of tibial spur, scutellary pore, punctured and depressed pronotal hind angles and distinctly bicoloured legs distinguish it from other common, similarly sized species such as *A (s.str.) aenea* (3).

2. *Amara strenua* Zimmermann

Length 8-9.6 mm. Black with a uniform metallic brassy or coppery reflection. Antennae black with segments 1-3 and base of fourth red; legs black with tibiae brown except apically. Pronotal front angles only slightly protruding forwards; sides almost parallel, base not punctured; inner foveae very deep and sharp, outer foveae very shallow, hind angles not depressed. Elytral striae rather fine, finely punctured, intervals convex posteriorly. Wings present. In coastal grassland, also coastal flood refuse; vi-x. Extremely local in south east England, Somerset and Suffolk; very scarce.
Similar species: See *A. plebeja* (1). Separated from similarly sized species also with a scutellary pore such as *A. (s.str.) ovata* (15) and *A. similata* (16), by the three-pronged tibial spur.

2. Subgenus *Amara sensu stricto*

This major part of the genus contains about 50 species, with 16 in Britain, ten in Ireland. They have a simple tibial spur (Fig. 210), antennae darkened from at least the fourth segment, and the prosternal process has no setae (Fig. 207). Some species have a scutellary pore.

Key to species of *Amara sensu stricto*

1. Legs entirely pale 2

- At least femora darkened, often dark brown or black 4

2. Scutellary pore present (Fig. 216) .... 4. *anthobia* Villa & Villa

- Scutellary pore absent 3

3. Front angles of pronotum strongly protruding (Fig. 217 – view from directly above) .... 10. *familiaris* (Duftschmid)

- Front angles of pronotum hardly protruding, so that front margin is almost straight (Fig. 218 – view from directly above) .... 11. *lucida* (Duftschmid)

4. Third antennal segment entirely darkened 5

- At least basal half of third antennal segment pale 8

5. Elytral striae fine to elytral apex where intervals are flat 6

- Elytral striae deepened apically where elytral intervals are convex 7

6. First and second antennal segments pale .... 17. *spreta* Dejean

- Second antennal segment darkened, only first segment pale .... 9. *famelica* Zimmermann

7. Length more than 7.5 mm; apical antennal segments at least twice as long as wide (Fig. 219) .... 12. *lunicollis* Schiödte

- Length less than 7.5 mm; apical antennal segments less than twice as long as wide (Fig. 220) .... 7. *curta* Dejean

8. Length less than 6 mm; scutellary stria absent or reduced to a few streaks (Fig. 221), scutellary pore absent; pronotum with four small, deep foveae (Fig. 221) .... 18. *tibialis* (Paykull)

- Length at least 6 mm; scutellary stria present (Fig. 222), sometimes also with a scutellary pore; pronotum with foveae not as above 9

9. Scutellary pore present (as in Fig. 216, opposite) 10

- Scutellary pore absent 14

10. Length more than 9.5 mm; inner pronotal foveae sharp and elongate (Fig. 222); apex of elytra strongly produced (Fig. 223) .... 8. *eurynota* (Panzer)

- Length 9 mm or less; inner pronotal foveae shallower, often indistinct; elytral apices hardly produced 11

11. Pore at hind angle of pronotum separated from side margin by about twice its diameter or less (Fig. 224 – view by tilting specimen slightly sideways) 12

- Pore at hind angle of pronotum separated from side margin by three to four times its diameter (Fig. 225 – view by tilting specimen slightly) 13

12. Base of pronotum not punctured; apex of aedeagus shorter in dorsal view (Fig. 226); tibiae as dark as femora, usually black .... 15. *ovata* (Fabricius)

- Base of pronotum at least partly punctured; apex of aedeagus longer in dorsal view (Fig. 227); tibiae often paler than femora .... 16. *similata* (Gyllenhal)

13. Front angles of pronotum angular, very strongly produced forwards and downwards (Fig. 228); pronotal foveae almost absent .... 13. *montivaga* Sturm

- Front angles of pronotum moderately produced but rounded at the tips (Fig. 229); inner pronotal foveae very fine and shallow but distinct .... 14. *nitida* Sturm

14. Elytral striae fine to elytral apex, where intervals are flat; inner pronotal fovea consisting of a short, deep streak (Plate 104, Fig. 230) .... 3. *aenea* (De Geer)

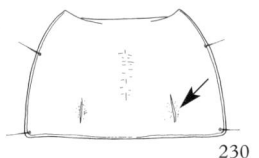

- Elytral striae deepened apically, where elytral intervals are convex; inner pronotal fovea shallow or more gradually defined 15

230

15. Base of pronotum punctured around foveae and towards hind angles (Fig. 231) 16

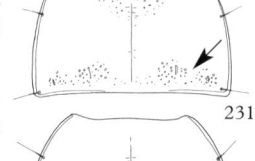

- Base of pronotum hardly or not punctured (Fig. 232) .... 14. *nitida* Sturm

231

232

16. Marginal row of elytral punctures with a wide interruption in the middle (side view Fig. 233); apical half of third antennal segment darkened; apex of aedeagus shorter and broader in dorsal view (Fig. 234) .... 5. *communis* (Panzer)

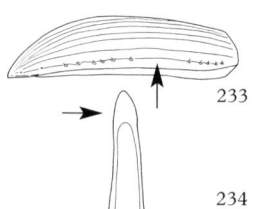

233

234

- Marginal row of elytral punctures more or less evenly spaced (side view Fig. 235); third antennal segment usually pale, at most darkened apically; apex of aedeagus longer and more slender in dorsal view (Fig. 236) .... 6. *convexior* Stephens

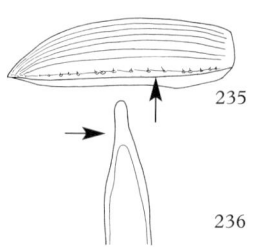

235

236

3. *Amara aenea* (De Geer) Plate 104

Length 6.5-8.8 mm. Shining brassy or coppery, rarely greenish, females with strong elytral microsculpture. Antennae black with three pale basal segments; legs black with tibiae mid to dark brown. Pronotum slightly narrower than elytra, evenly narrowed in front of the sharp hind angles; base unpunctured, inner foveae sharp, deep and elongate (Fig. 230). Scutellary pore absent; striae very fine, with flat intervals; elytral apex produced and sides sinuate sub-apically, especially in females. Wings present. In dry grasslands, gardens, dunes and waste land; iv-viii. Widespread in Britain and Ireland; very abundant.

Similar species: The fine elytral striae and extended elytral apex resemble *A. spreta* (17) but that species has only two pale antennal segments and the base of the pronotum more punctured. The fine striae and short, deep pronotal fovea without surrounding punctures distinguish *A. aenea* from *A. communis* (5) and *A. convexior* (6). Smaller than *A. eurynota* (8) which has similar pronotal foveae and pointed elytral apices, and also lacking the scutellary pore of that species. The three pale basal antennal segments separate it from *A. famelica* (9) and *A. spreta* (17).

4. *Amara anthobia* Villa & Villa

Length 5.6-7 mm. Black with a coppery or brassy reflection. Antennae dark with basal three or four segments pale; legs entirely pale brown. Pronotum narrower than elytra, front margin almost straight, sides slightly rounded to hind angles, base unpunctured, inner foveae distinct. Elytra with sutural stria distinct, bent outwards basally to meet the scutellary stria; scutellary pore present (Fig. 216); intervals flat basally, convex apically. Wings present. In open, sandy sites and coastal dunes; iv-vi. Local in England as far north as Lancashire; usually scarce.

Similar species: Resembles *A. lucida* (11) but that species has no scutellary pore; this also separates it from the larger *A. familiaris* (10) which has more protruding pronotal front angles.

5. *Amara communis* (Panzer)

Length 6-8 mm. Shining brassy black, rather parallel-sided. Antennae with basal 2½ segments pale, remainder black; legs black with tibiae slightly paler. Pronotum rather long in proportion to the elytra; front angles protruding strongly forwards and downwards; foveae shallow but distinct, base punctured except in the foveae (Fig. 231). Scutellary pore absent; elytral striae deepened apically, where intervals are convex; elytral margin with the row of pores widely interrupted in the middle (Fig. 233) (sometimes with a single puncture in this gap on one side). Aedeagal apex short and symmetrically tapered (Fig. 234). Wings present. In grasslands and moorland, even if wet; iv-vii. Widespread in Britain and Ireland; very abundant.

Similar species: Separated from the slightly larger *A. convexior* (6) by the interrupted marginal row of pores on the elytra, and usually also by the apical part of the third antennal segment being darkened. The long pronotum resembles *A. lunicollis* (12) but in that species the third antennal segment is completely darkened, and the pronotum is even wider and more rounded in front of the hind angles. The punctures around the pronotal foveae distinguish both this species and the next from aberrant *A. nitida* (14) that lack the usual scutellary pore. See also *A. aenea* (3).

6. *Amara convexior* Stephens

Length 6.5-8.2 mm. Black with a slight metallic reflection. Antennae dark with three pale basal segments, apex of third segment rarely slightly darkened; legs black with base of tibiae brown. Pronotal sides rounded, base quite extensively punctured. Scutellary pore absent; elytral striae deepened apically, where intervals are convex; elytral margin with the row of pores more widely spaced but not clearly interrupted in the middle (Fig. 235). Aedeagal apex long and asymmetrical (Fig. 236). Wings present. In dry, well drained sites with ruderal vegetation; iv-viii. Widespread in south-east England, local further north and west, absent from Scotland and Ireland; usually scarce.

Similar species: See *A. aenea* (3) and *A. communis* (5).

7. *Amara curta* Dejean

Length 5.5-7.2 mm. Black, rather convex, with a slight metallic reflection. Appendages black with tibiae slightly paler; antennal segments 1-2 pale, third segment mostly darkened; antennae short, apical segments less than twice as long as wide (Fig. 220). Pronotal sides parallel or very slightly contracted in front of the sharp hind angles; foveae shallow but distinct, sometimes slightly punctured. Scutellary pore absent; elytra short, not produced apically, where striae are convex. Wings present. In dry grasslands, heath

and dunes; iv-vii. Widespread but very local in England, with only a single recent Scottish record; very scarce.
Similar species: Smaller and with shorter antennae than *A lunicollis* (12) and other species with dark third antennal segment; separated from similarly sized *A. anthobia* (4), *A. familiaris* (10) and *A. lucida* (11) by the darkened femora and third antennal segment.

8. *Amara eurynota* (Panzer)

Length 9.6-12.5 mm. Metallic coppery or bronze, with strong microsculpture. Appendages black except for paler basal three antennal segments. Pronotum evenly rounded in front of the slightly acute hind angles, hind margin slightly sinuate; base not punctured, inner foveae deep, sharp and elongate (Fig. 222), outer foveae very faint; pore inside hind angles separated from hind angle by about twice its diameter. Scutellary pore present; elytral striae fine but intervals convex and slightly ridged, giving the elytra a corrugated appearance when lit from the side; apices produced, margins sinuate sub-apically (Fig. 223). Wings present. In arable fields, dunes and other open, rather dry situations; vi-x. Widespread in England and Wales, very local in Scotland and Ireland; sometimes abundant.
Similar species: Larger than all other species of the genus. Separated from *A. ovata* (15) and *A. similata* (16) by its larger size and deep pronotal fovea. See also *A. aenea* (3).

9. *Amara famelica* Zimmermann

Length 6.7-9 mm. Black, with a bronze or greenish reflection. Appendages black, first antennal segment slightly paler. Pronotal sides parallel-sided in front of hind angles, base not punctured, inner foveae linear, outer foveae reduced to a small pit. Scutellary pore absent; elytral striae fine, intervals flat, apices protruding and margins sinuate sub-apically. Wings present. On sandy heaths; iv-v. Extremely local in southern and eastern England as far north as Yorkshire; extremely scarce.
Similar species: Resembles *A. spreta* (17) in the fine elytral striae and flat intervals but separated by having only the first antennal segment at all pale. Elytral striae finer and intervals flatter than in *A. lunicollis* (12). See also *A. aenea* (3).

10. *Amara familiaris* (Duftschmid)

Length 5.5-7.3 mm. Dark metallic brassy or coppery. Antennae dark with basal three and part of fourth segments pale; legs entirely pale to mid reddish brown. Eyes rather flat. Front angles of pronotum protruding and angulate (Fig. 217); sides slightly rounded in front of hind angles, base not punctured, inner foveae small, sometimes with a transverse depression joining them across the mid-line. Scutellary stria distinct, often reaching elytral base, scutellary pore absent; intervals convex apically. Wings present. In open grasslands, heaths and dunes; iv-vii. Widespread in Britain, local in Ireland; abundant.
Similar species: See *A. anthobia* (4); that species and *A. lucida* (11) are both smaller with less protruding pronotal front angles (which must be viewed from directly above to appreciate the difference).

11. *Amara lucida* (Duftschmid)

Length 4.7-6.5 mm. Black with a greenish or brassy metallic reflection. Antennae dark with basal three or four segments pale; legs entirely pale. Front margin of pronotum almost straight (Fig. 218), sides parallel in front of hind angles; base unpunctured, inner

foveae very small and shallow. Elytra with scutellary stria fine, not reaching elytral base, scutellary pore absent; intervals convex apically. Wings present. In sand dunes and dry grassland, mostly coastal; iv-vii. Local around the coast of England and inland in Norfolk and Suffolk; very local elsewhere in England; extremely local in Ireland; sometimes abundant.
Similar species: See *A. anthobia* (4) and *A. familiaris* (10).

12. *Amara lunicollis* Schiödte

Length 7.5-9 mm. Convex with pronotum notably wider than elytra; shining black or with a coppery reflection. Appendages black except first one or two antennal segments; apical segments of antennae more than twice as long as wide (Fig. 219). Pronotal sides strongly rounded and contracted in front of the blunt hind angles; foveae distinct, usually linear, outer foveae oblique, base unpunctured. Scutellary pore absent; elytral intervals convex, especially apically; elytral apices not produced. Wings present. In most open or semi-open habitats, especially if well drained yet not too dry; v-viii. Widespread in Britain and Ireland; often abundant.
Similar species: See *A. communis* (5), *A. curta* (7) and *A. famelica* (9). More convex than these and *A. spreta* (17) which also has the darkened third antennal segment.

13. *Amara montivaga* Sturm

Length 7.8-9.1 mm. Black or metallic coppery or brassy. Appendages black except for basal 3½ antennal segments. Pronotum very large and wide, especially in males, front angles protruding strongly forwards and almost angular (Fig. 228); dorsal surface evenly rounded and almost without foveae or punctures; pore inside hind angle separated from the side margin by at least four times its diameter (Fig. 225). Scutellary pore present; elytral intervals slightly convex throughout, apices not produced. Wings present. In open, sandy or chalky sites with ruderal vegetation; v-viii. Local in the southern half of England and Wales; in Ireland known only from Kerry; usually scarce.
Similar species: Similar to *A. nitida* (14) but with sharper pronotal hind angles and darker tibiae. The large, evenly convex pronotum with almost complete lack of foveae also distinguishes *A. montivaga* from *A. ovata* (15) and *A. similata* (16) in which the pore at the pronotal hind angle is much closer to the side margin.

14. *Amara nitida* Sturm

Length 7.2-8.5 mm. Metallic brassy or coppery. Antennae black with basal three segments pale; legs black or dark brown with pale tibiae. Pronotum with front angles moderately protruding but rounded (Fig. 229); foveae shallow but distinct, with at most a few punctures between foveae and basal margin; pore inside hind angle separated from the side margin by three to four times its diameter. Scutellary pore usually present at least on one side but sometimes completely missing; elytral intervals convex apically, elytral apex not produced. Wings present. In dry, sometimes shaded habitats on well drained soils; v-vii. Extremely local throughout England and Wales; very scarce.
Similar species: See *A. communis* (5), *A. convexior* (6) and *A. montivaga* (13). The pore at the pronotal hind angle is further from the side margin than in *A. ovata* (15) and *A. similata* (16).

15. *Amara ovata* (Fabricius)

Length 8-9.5 mm. Black, often with a brassy reflection. Antennae black with basal three segments pale; legs black. Pronotum as wide as elytra, evenly rounded in front of hind angles; foveae shallow, not punctured; pore at hind angles separated from side margin by twice its diameter (Fig. 224). Scutellary pore present; elytral striae deep, intervals convex apically, apices hardly produced. Apical part of aedeagus as wide as long (Fig. 226). Wings present. In open, dry fields and gardens; v-viii. Widespread in Britain and Ireland; abundant.
Similar species: Closely related to *A. similata* (16) but more robust, with a wider pronotum relative to the elytra, no punctures around the pronotal foveae and tibiae not paler than the femora. The aedeagal apex is wider than that of *A. similata*. See also *A. eurynota* (8), *A. montivaga* (13) and *A. nitida* (14).

16. *Amara similata* (Gyllenhal)

Length 8-9.5 mm. Metallic brassy or coppery. Antennae black with basal three segments pale; legs black with tibiae often mid to dark brown. Pronotum narrower than elytra, strongly contracted in front and hardly rounded in front of hind angles; base usually punctured at least around the shallow foveae; pore at hind angles separated from side margin by twice its diameter. Scutellary pore present; elytral striae deep, intervals convex apically, apices hardly produced. Apical part of aedeagus longer than wide (Fig. 227). Wings present. In open fields and gardens, often near water; v-viii. Widespread in England and Wales, local in Scotland and Ireland; abundant.
Similar species: See *A. eurynota* (8), *A. montivaga* (13) *A. nitida* (14) and *A. ovata* (15).

17. *Amara spreta* Dejean

Length 7.5-9.5 mm. Metallic bronze, with a reddish or purple reflection, strongly microsculptured. Appendages black with tibiae and basal two antennal segments pale. Pronotum widened almost to the sharp hind angles, hind margin slightly sinuate, base punctured, foveae sharp and often deep. Scutellary pore absent; elytral striae fine and intervals flat, apices produced so that lateral margins are sinuate sub-apically. Wings present. In sand pits and dunes; iv-vi, ix-x. Extremely local in England and south Wales; can be abundant.
Similar species: See *A. aenea* (3), *A. famelica* (9) and *A. lunicollis* (12).

18. *Amara tibialis* (Paykull)

Length 4.5-5.9 mm. Body very parallel-sided, black often with a brassy metallic reflection. Antennae black with segments one to three pale (very rarely third segment darkened), apical segments parallel-sided and elongate; legs nearly black with brown tibiae. Pronotum slightly contracted to sharp hind angles, base straight, unpunctured, with four deep but short foveae near hind margin (Fig. 221). Elytra almost parallel-sided, scutellary pore absent, scutellary stria usually reduced to a few punctures or completely lacking; striae very fine, sometimes clearly punctured, intervals flat. Wings present. In sand pits, dry heaths, dunes and well drained open ground; iv-viii. Widespread in southern and eastern England; mainly coastal further north, in Wales and in Ireland; often very abundant.
Similar species: Smaller than all other species except *A. (Celia) infima* (21) which also has a reduced scutellary stria but is more rounded, with the apical antennal segments short, rounded and not entirely black. The reduced scutellary stria, characteristic pronotal foveae and darkened femora distinguish *A. tibialis* from small examples of *A. lucida* (11) and the first two of these features, plus the pale third antennal segment, separate it from *A. curta* (7).

3. Subgenus *Celia* Zimmermann

This group is characterised by the pale antennae and lack of ridge outside the pronotal foveae; the hind angles are not or hardly toothed and the prosternal process has no setae. It is a large and rather diverse assemblage, with only four British species (two in Ireland). Two specimens of *A (Celia) cursitans* (Zimmerman) were recorded from London in 1956, but it has not become established.

Key to species of *Amara* (*Celia*)

1. Scutellary stria absent or very short; length usually less than 5.5 mm
   .... 21. *infima* (Duftschmid)

- Scutellary stria fully developed; length usually more than 5.5 mm
   2

2. Scutellary pore present (as in Fig. 216) ....
   .... 22. *praetermissa* (Sahlberg, C.R)

- Scutellary pore absent 3

3. Base of pronotum extensively punctured (Plate 105) ....
   .... 19. *bifrons* (Gyllenhal)

- Base of pronotum punctured in foveae only 4

4. Antennae entirely pale; hind angles of pronotum not projecting
   .... 20. *fusca* Dejean

- Antennae slightly darkened from fifth segment to apex; hind angles of pronotum slightly projecting .... [*cursitans* (Zimmerman)]

19. *Amara bifrons* (Gyllenhal) Plate 105

Length 5.5-7.3 mm. Mid to dark brown with a brassy metallic reflection. Appendages pale yellow-brown. Pronotum no wider than elytra, sides evenly rounded, hind angles with a suggested minute tooth, foveae shallow, base strongly and evenly punctured. Scutellary pore absent; sides of elytra slightly rounded, striae fine and hardly punctured, intervals convex apically. Wings present. On open sites on well drained soils; v-ix. Widespread in Britain, very local in Ireland; sometimes abundant.
Similar species: Smaller than both *A. praetermissa* (22) and *A. fusca* (20) and with the pronotal base more uniformly punctured that *A. fusca*. This pronotal punctuation also distinguishes *A. bifrons* from immature, pale examples of most *Amara s.str.* species.

20. *Amara fusca* Dejean

Length 7.5-8.5 mm. Mid to dark brown, with a coppery reflection, elytra sometimes reddish. Appendages uniformly mid to light brown. Pronotum narrower than elytra, sides rounded, especially just in front of hind angles, which are minutely toothed; hind margin slightly sinuate, foveae distinct and punctured but without ridge inside the hind angles. Scutellary pore absent, elytral striae narrow and finely punctured, intervals almost flat. Wings present. In dry, open sandy places such as sand and gravel workings, roadside verges and dunes; viii-x. Only known recently from Suffolk, although formerly also in Glamorgan, Kent and Sussex; scarce.

Similar species: Larger than other species of *Celia* and most easily confused with *A. (Bradytus) consularis* (25) and *A. (Percosia) equestris* (27), both of similar size and with pale antennae. *A. fusca* differs from these species in lacking any distinct oblique ridge delimiting the outer pronotal fovea and separating it from the hind angle. Also the eyes are more convex than *A. equestris* and the head narrower than *A. consularis*.

21. *Amara infima* (Duftschmid)

Length 4.9-5.6 mm. A short, rounded and convex species; black or dark, slightly metallic brown, shining. Antennae very short, apical segments quadrate with rounded sides; brown or with segments two to four (rarely whole of apex) somewhat darkened but not abruptly black. Pronotum with front margin almost straight, sides evenly rounded, foveae small but deep and slightly punctured. Elytra with slightly bulging, rounded sides; scutellary stria lacking or very short; striae punctured, intervals flat except apically. Wings usually absent. On sandy heaths and dunes; ix-iv. Extremely local in southern England, East Anglia and south Wales; scarce.

Similar species: See *A. (s.str.) tibialis* (18).

22. *Amara praetermissa* (Sahlberg, C.R)

Length 6.2-8 mm. Dark brown, hardly metallic. Appendages mid to light brown. Pronotum wide and convex, as wide as elytra, sides evenly rounded, base extensively punctured except medially, foveae shallow. Scutellary pore present; elytral striae deep and clearly punctured, intervals convex. Wings present. In open, dry, well drained and often disturbed habitats, including grassland, dunes, and 'brownfield' sites; vii-x. Local in eastern England, and on the coasts of Wales, southern Scotland and southern Ireland; usually scarce.

Similar species: Distinguished from all other *Celia* by the presence of a scutellary pore. The wide pronotum with extensive basal punctuation separates *A. praetermissa* from immature, pale specimens of *Amara s.str.* that may also have a scutellary pore.

4. Subgenus *Paracelia* Bedel

There is a single British species in this Holarctic subgenus of more than 20 species, characterised by having setae on the prosternal process (Fig. 211), together with flat eyes (Fig. 215) and fine elytral striae with flat intervals.

23. *Amara quenseli* (Schönherr) Plate 106

Length 6.5-9.5 mm. A rather variable species: mid to dark brown, often with a metallic reflection, sometimes reddish. Antennae pale basally, usually somewhat gradually darkened from the third to fifth segment to the apex; legs brown, femora and tibiae sometimes darker. Eyes small and flat. Pronotum narrower than elytra, sides rounded or almost straight in front of the hind angles, which are sometimes minutely protruding; rear part of side margin obliquely depressed; foveae distinct, base punctured at least around foveae, sometimes completely. Prosternal process with two to four setae. Elytra large, scutellary pore absent; striae fine and intervals flat; apex not acuminate. Wings usually present. In open sandy and gravelly deposits often near rivers; vii-x. Very local in

north-east Scotland and the Isle of Rum; usually scarce.
Similar species: Resembles the larger *Celia* species and *A. (Bradytus) consularis* (25) but can be recognised by the combination of rather flat eyes, flat elytral intervals and lack of ridge outside the outer pronotal fovea. Specimens with darkened antennae are distinguished from *Amara s.str.* species by the flat eyes, and the obliquely depressed pronotum in front of the hind angles.

5. Subgenus *Bradytus* Stephens

This subgenus is characterised by the combination of pale antennae and the presence of a ridge between the outer pronotal foveae and the hind angles, which are either protruding or slightly toothed (Fig. 238). The head is also wider relative to the pronotum than in the other British subgenera. There are over 50 Holarctic species; three autumn-breeding species occur in both Britain and Ireland.

Key to species of *Amara* (*Bradytus*)

1. Colour pale yellow-brown; pronotal sides and hind margin sinuate, so that hind angles are sharp and acute (Fig. 237) .... 26. *fulva* (Müller, O.F)

- Colour mid to dark brown; pronotal hind angles more or less right-angled, with slight protruding tooth (Figs 238, 239) 2

2. Ridge inside pronotal hind angle less oblique, with pore situated between this and the hind angle (Fig. 238); elytra widened just behind front angles, which are therefore obtuse (Plate 107) .... 24. *apricaria* (Paykull)

- Ridge inside pronotal hind angle more oblique, with pore situated at its base (Fig. 239); elytral sides almost straight behind front angles, which are more or less right-angled (Fig. 239) .... 25. *consularis* (Duftschmid)

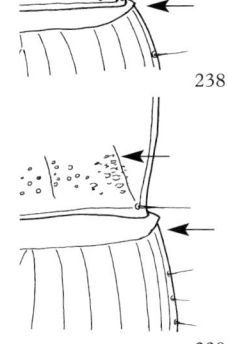

24. *Amara apricaria* (Paykull) Plate 107

Length 6.5-8.5 mm. Mid to dark reddish brown, with a metallic reflection. Appendages mid brown. Pronotum very transverse, widest at middle of sides; behind this narrowed to the slightly toothed hind angles; base punctured except near mid-line, outer foveae delimited externally by a slightly oblique ridge; pore situated between the base of this ridge and the hind angle (Fig. 238). Elytra widened just behind the strong shoulder tooth, then almost parallel-sided to near apex; striae strongly punctured, intervals convex. Wings present. On open, often cultivated land, and where there is ruderal vegetation; vi-ix. Widespread in eastern England and Scotland, local further west and in Ireland; abundant.
Similar species: Smaller and more parallel-sided than *A. consularis* (25). Separated from similarly-sized species with pale antennae and punctured pronotal base such as *A. (Celia) bifrons* (19) and *A. (Paracelia) quenseli* (23) by the distinct oblique ridge inside the slightly toothed pronotal hind angles.

25. *Amara consularis* (Duftschmid)

Length 8-9.5 mm. Reddish-brown to black, elytra sometimes paler than head and pronotum, hardly metallic. Appendages red-brown. Pronotal sides rounded anteriorly but little contracted to the hind angles; base punctured around the foveae and sometimes also medially; ridge inside hind angle clearly oblique and ending basally in a pore (Fig. 239). Elytra not widened just behind shoulders, sides evenly rounded; striae moderately punctured, intervals slightly convex. Wings present. In dry sand and gravel pits with vegetation, also lowland sandy heaths; viii-xii. Local in England; recorded from Ireland before 1900 only; usually scarce.
Similar species: See *A. apricaria* (24) and *A. (Celia) fusca* (20). Also resembles *A. (Percosia) equestris* (27) in size and pronotal features but has a narrower pronotal side border, more convex eyes and no setae on the prosternal process.

26. *Amara fulva* (Müller, O.F)

Length 8.2-10.5 mm. Pale yellowish-brown, often with a green metallic reflection. Appendages yellow. Pronotum very transverse, widest medially; sides contracted and sinuate behind this; base also sinuate so that hind angles are sharp and acute (Fig. 237); foveae punctured; ridge outside outer foveae very oblique, with a pore basally. Elytral striae deep and finely punctured, intervals convex. Wings present. In sandy and gravelly sites often near water, as well as dunes; vii-x. Very local in Britain and Ireland; sometimes abundant.
Similar species: Distinguished from mature specimens of all other *Amara* species by the pale colour. The sinuate pronotal sides and acute hind angles will separate it from immature (pale) examples of other similarly sized species.

6. Subgenus *Percosia* Zimmermann

This small subgenus, characterised by the multiple setae on the prosternal process, has a single species in Britain.

27. *Amara equestris* (Duftschmid) Plate 108

Length 7.5-10.5 mm. Rather convex, non metallic dark brown to black, margins of pronotum paler. Appendages mid brown. Eyes large (Fig. 214). Side borders of pronotum thick, especially towards hind angles, which have an oblique ridge inside but are not toothed; foveae (and sometimes all of base) punctured. Prosternal process with 6 or more setae. Elytral striae rather finely punctured, intervals distinctly convex apically. Wings present. In dry, sandy places, dunes and gravel pits; vii-x. Extremely local in Britain; very scarce.
Similar species: See *A. (Bradytus) consularis* (25). Superficially also resembles some *Harpalus* species because of the convex form and thickened pronotal side borders.

3. *CURTONOTUS* Stephens

A group of about 80 Holarctic species, previously considered as part of *Amara* but distinguished by the unbordered prosternal process (Fig. 208), as well as by their larger size and sinuate side margins of the pronotum. There are three species in Britain, two of which also occur in Ireland.

Key to species of *Curtonotus*

1. Antennae more or less uniformly pale to mid brown; length usually more than 11 mm 2

- Antennae darkened, usually from fourth segment to apex; length less than 11 mm .... 1. *alpinus* (Paykull)

2. Raised side border of pronotum not reaching hind angles which are strongly protruding (Fig. 240); elytra shorter, together 1.5 times as long as wide (Plate 109) .... 2. *aulicus* (Panzer)

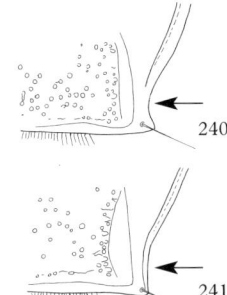

- Raised side border of pronotum extending back to non-protruding hind angles (Fig. 241); elytra longer, together 1.7 times as long as wide .... 3. *convexiusculus* (Marsham)

1. *Curtonotus alpinus* (Paykull)

Length 8-10.5 mm. Black, sometimes with reddish elytra, non-metallic. Antennae with basal three segments reddish, remainder gradually darkened apically, often mostly black; legs all black, or with femora and part of tibiae reddish. Sides of pronotum rounded, contracted basally and sinuate only just in front of the minute hind angles; side borders narrow but distinct throughout; foveae deep, base punctured except medially, ridge outside foveae meeting pronotal base. Elytra parallel-sided, together about 1.6 times as long as wide; striae finely punctured, intervals almost flat. Wings usually absent. On dry moraines and mountain tops; vi-x. Only in the Scottish Highlands, very scarce.
Similar species: Intermediate in shape between the other two species of the genus but smaller than both, with darkened antennae and much less basally sinuate pronotal sides. Superficially resembles some *Pterostichus* species but these do not have the multiple setae on the penultimate segment of the labial palpi that are found in *Curtonotus*.

2. *Curtonotus aulicus* (Panzer) Plate 109

Length 11-14 mm. Black or dark reddish brown, often with a brassy metallic reflection. Appendages mid brown, except apices of tibiae sometimes darker. Pronotum very transverse, strongly contracted and sinuate in front of the strongly protruding hind angles; side borders wide, and ending in front of hind angles where they meet the ridge outside the foveae (Fig. 240); pronotal base densely punctured throughout. Elytra very convex, together only a little over 1.5 times as long as wide; intervals convex. Wings present. In almost all open, dry habitats where there is herbaceous vegetation in seed; vii-x. Widespread in Britain and Ireland; very abundant.
Similar species: More convex and wider than *C. convexiusculus* (3) with the raised side borders of the pronotum not reaching the hind angles.

3. *Curtonotus convexiusculus* (Marsham)

Length 11-12.8 mm. Slightly metallic dark brown or bronze. Appendages entirely mid to light brown; antennae long, especially the slender basal three segments. Pronotum strongly contracted basally and sinuate a little in front of the acute hind angles; side borders narrow but distinct throughout (Fig. 241); foveae deep, base punctured, ridge

outside foveae meeting pronotal base. Elytra long and parallel-sided, together at least 1.7 times as long as wide; striae finely punctured, intervals convex. Wings present. On coastal saltmarshes, dunes and lowland grasslands; vii-x. Widespread around the east and south coasts of England and Wales, local inland in eastern England and in Scotland; in Ireland only from Waterford; abundant.
Similar species: See *C. aulicus* (2).

Tribe HARPALINI

This worldwide tribe of over 2000 species is distinguished from the Pterostichini and Platynini by having only a single setiferous puncture inside each eye, no seta at the pronotal hind angle and by the pubescent third antennal segment. The species are variable in size and also in feeding habits. Several are seed feeders; others are omnivorous. There are 11 British genera (nine in Ireland); *Ophonus* was previously considered as a subgenus of the major genus *Harpalus* and *Anthracus* was part of *Acupalpus*. Some recent authors include *Trichocellus* within *Dicheirotrichus*. The genera *Diachromus* and *Scybalicus* (with a single species each) are only occasional immigrants in Britain but are included as both have occurred recently.

Key to genera of Harpalini

1. Scutellary stria of elytra reduced or absent (Plates 116, 117); at least outer elytral intervals punctured and pubescent 2

- Scutellary stria present; or if somewhat reduced, elytra not at all pubescent 3

2. Upper surface entirely pubescent (Plate 116); length usually more than 5.5 mm .... 6. *Dicheirotrichus* Jacquelin du Val (p. 167)

- Only outer intervals of elytra pubescent (Plate 117); length usually less than 5.5 mm .... 7. *Trichocellus* Ganglbauer (p. 168)

3. Elytra entirely pubescent 4

- Elytra glabrous, or pubescent only at sides (and sometimes also apically) 7

4. Reddish brown with metallic blue-green pronotum and apical mark on elytra .... 4. *Diachromus* Erichson (p. 167)

- Not coloured as above, usually uniformly coloured 5

5. Head punctured and pubescent, sometimes sparsely 6

- Head glabrous, except for setae inside the eyes (Plate 110) ....
.... 1. *Harpalus* Latreille (part) (p. 149)

6. Elytral base depressed and basal margin sinuate in front of stria 3 (Fig. 242) .... 5. *Scybalicus* Schaum (p. 167)

- Elytral base level, basal margin straight or nearly so (Plates 113, 114) .... 2. *Ophonus* Dejean (p. 158)

7. Pronotum with distinct and complete basal border (Fig. 243); length usually more than 6 mm 8

- Pronotum without basal border or it is present only laterally; length usually less than 6 mm 9

8. Apical spur of hind tibiae as long as or longer than first hind tarsal segment (Fig. 244) .... 1. *Harpalus* Latreille (part) (p. 149)

- Apical spur of hind tibiae shorter than first hind tarsal segment (Fig. 245) .... 3. *Anisodactylus* Dejean (p. 165)

9. Antennae entirely pale or slightly darkened from fourth segment to apex; mentum toothed medially (Fig. 246) .... 8. *Bradycellus* Erichson (p. 169)

- Antennae darkened, usually from the second or third segment to apex; mentum not toothed 10

10. Length at least 5 mm .... 9. *Stenolophus* Dejean (p. 172)

- Length less than 5 mm 11

11. Pronotal hind angles rounded, sides not sinuate (Plate 120) .... 10. *Acupalpus* Latreille (p. 173)

- Pronotal sides sinuate in front of sharp hind angles (Plate 121) .... 11. *Anthracus* Motschulsky (p. 177)

1. HARPALUS Latreille

A large genus of almost 500 species, with some 150 in Europe and 20 in Britain (but only eight in Ireland). They are characterised by the medium to large size (excepting *H. pumilus* Sturm); a long apical spur on the hind tibiae (Fig. 244); sutural stria present, and the elytra usually non-pubescent excepting the subgenus *Pseudoophonus*. *Ophonus*, often treated as a subgenus of *Harpalus*, is considered here as a separate genus. Two species are included in the keys only: *H. (s.str.) cupreus* Dejean occurred on the Isle of Wight until 1914; *H. (Pardileus) calceatus* (Duftschmid) is an occasional migrant that has been found in Sussex,

Essex and Yorkshire. In most species the females are much duller than the males, with coarser microsculpture. The three subgenera are sometimes given generic status and are keyed separately for convenience.

Key to subgenera of *Harpalus*

1. Tarsi with fine dorsal pubescence (view mid tarsi – Fig. 247) 2

- Tarsi with lateral and ventral setae but no dorsal pubescence .... 2. *Harpalus sensu stricto* (p. 151)

2. Elytra wholly or partly pubescent .... 1. *Pseudoophonus* Motschulsky (p. 150)

- Elytra entirely non pubescent .... 3. *Cryptophonus* Brandmayr & Zetto Brandmayr (p. 158)

1. Subgenus *Pseudoophonus* Motschulsky

This Holarctic group has more than 80 species, many in Asia and north America. They are distinguished by the combination of pubescent tarsi and wholly or partly pubescent elytra, as well as a densely punctured pronotal base. *H. (Pardileus) calceatus*, sometimes also included in *Pseudoophonus*, is included in the key.

1. All elytral intervals entirely punctured and pubescent 2

- Only outer intervals finely punctured and pubescent .... [*calceatus* (Duftschmid)]

2. Sides of pronotum slightly sinuate in front of acute hind angles (Plate 110); length at least 11 mm .... 2. *rufipes* (De Geer)

- Sides of pronotum straight and slightly contracted in front of obtuse hind angles (Fig. 248); length usually less than 11 mm .... 1. *griseus* (Panzer)

1. *Harpalus griseus* (Panzer)

Length 8-12 mm. Black, appendages yellow-brown. Pronotal sides behind middle straight and slightly contracted to obtuse hind angles; base uniformly punctured but hardly rugose. Elytral intervals uniformly pubescent. Underside of abdomen pubescent medially, almost glabrous at sides. Wings present. Only known from recent captures at light traps in south London; possibly migrant and not established.

Similar species: Smaller than *H. rufipes* (2), the only other *Harpalus* with wholly pubescent elytra, with sides of pronotum less sinuate in front of the blunter hind angles, and differing pattern of pubescence beneath the abdomen.

2. *Harpalus rufipes* (De Geer) Plate 110

Length 11-16 mm. Black, rather dull. Appendages reddish-brown, apices of femora sometimes darkened; tarsi with dorsal pubescence. Pronotum with sides rounded, and sinuate just in front of the slightly acute hind angles; base almost entirely punctured and rugose. Elytral intervals uniformly punctured and with short golden pubescence. Underside of abdomen smooth medially but with fine pubescence at sides. Wings present. In open, dry situations on light soils, especially arable fields; vi-ix. Widespread in Britain and Ireland except in the uplands; very abundant.

Similar species: See *H. griseus* (1). Distinguished from the larger *Ophonus* species by the non-pubescent head and pronotum, and the sharp pronotal hind angles.

2. Subgenus *Harpalus sensu stricto*

This subgenus comprises the majority of the British and Irish species, including *H. froelichii* sometimes separated into the subgenus *Haploharpalus* Schauberger. They all lack pubescence on the dorsal surface of the tarsi, and all except *H. affinis* have non-pubescent elytra.

Key to species of *Harpalus sensu stricto*

1. Outer intervals of elytra pubescent (Plate 111) .... 3. *affinis* (Schrank)

- Elytra not at all pubescent 2

2. Length less than 6.5 mm; elytra without scutellary pore (Fig. 249) .... 12. *pumilus* Sturm

- Length more than 6.5 mm; elytra with scutellary pore 3

3. Seventh elytral stria with a row of three or more sub-apical punctures (Fig. 250) 4

- Seventh elytral striae without, or at most two punctures on one side only 7

4. Length more than 12 mm; abdominal segments with fine pubescence .... 6. *dimidiatus* (Rossi)

- Length usually less than 12 mm; abdominal segments with only the usual pair of long setae 5

5. Antennae entirely pale .... 13. *rubripes* (Duftschmid)

- At least second and third segments of antennae dark 6

6. Upper surface black or with a slight metallic blue reflection .... 14. *rufipalpis* Sturm

- Upper surface distinctly metallic green or blue .... 8. *honestus* (Duftschmid)

7. Abdominal segments with fine pubescence (side view Fig. 251) 8

- Abdominal segments with only the usual pair of long setae 10

8. Base of pronotum extensively punctured (Fig. 252) .... 17. *smaragdinus* (Duftschmid)

- Base of pronotum unpunctured except in the narrow foveae 9

9. Antennae entirely pale .... 7. *froelichii* Sturm

- At least second and third segments of antennae darkened .... 11. *neglectus* Audinet-Serville

10. Tibiae brown, contrasting with the black femora; sides of pronotum contracted and sinuate in front of sharp hind angles, base punctured in and around the foveae only (Fig. 253) .... 5. *attenuatus* Stephens

- Tibiae and femora either both black or both pale; sides of pronotum straight or rounded, base punctured either extensively (Fig. 254) or hardly at all (Figs 257-259) 11

11. Base of pronotum extensively punctured or roughened (Fig. 254) 12

- Base of pronotum unpunctured and smooth except in the narrow foveae (Figs 257-259) 14

12. Colour bright metallic green or coppery .... [*cupreus* Dejean]

- Colour black or brown 13

13. Third elytral interval with two or three foveate punctures in apical third, close to stria 2 (Fig. 255) .... 9. *laevipes* Zetterstedt

255

- Third elytral interval with only a single shallow puncture (Fig. 256) .... 10. *latus* (Linnaeus)

256

14. Pronotum very transverse, 1.5 times as wide as long, base strongly concave (Fig. 257); body colour at least partly brown .... 16. *servus* (Duftschmid)

257

- Pronotum not as transverse, base straight or less concave (Figs 258, 259); body colour black 15

15. Length less than 8.5 mm; pronotal base slightly concave, hind angles sharp (Fig. 258) .... 4. *anxius* (Duftschmid)

258

- Length more than 8.5 mm; pronotal base straight, hind angles rounded (Fig. 259) 16

16. Antennae entirely pale .... 18. *tardus* (Panzer)

259

- At least antennal segments 2-4 darkened, often black .... 15. *serripes* (Quensel in Schönherr)

3. *Harpalus affinis* (Schrank) Plate 111

Length 9-12 mm. Extremely variable in colour; typically metallic golden-green but can be reddish copper, bluish purple or even black with only a hint of metallic reflection. Appendages pale or darkened, except first antennal segment always pale. Pronotum widest in front of middle, sides straight and contracted behind to the slightly rounded hind angles; base extensively punctured except sometimes medially (Fig. 243). Third elytral interval with one to three very fine punctures; outer three and apical intervals punctured and pubescent. Females with abrupt sub-apical sinuation of elytral margin. Wings present. In gardens, waste ground, arable fields and almost all open dry situations; v-ix. Widespread in Britain, local in the east of Ireland; very abundant.

Similar species: Distinguished from other metallic green species such as *H. honestus* (8) and *H. rubripes* (13) by the punctured and pubescent outer intervals of the elytra.

4. *Harpalus anxius* (Duftschmid)

Length 6.5-8 mm. Black, males moderately shining, females dull, especially elytra. Antennae pale, often with second and third segments darkened; legs except tarsi dark; front tibiae with three pre-apical spines. Pronotum widest in front of middle, sides almost straight and slightly contracted to sharp, rectangular hind angles; base usually unpunctured but sometimes rugose between the shallow narrow foveae; hind margin slightly concave (Fig. 258). Elytral striae fine, unpunctured, intervals flat except in apical third. Wings present. On dunes and inland sand pits and sandy heaths; v-viii. Widespread around the coast of England and Wales from Lincolnshire to Cumbria: also Isle of Man and the south-east coast of Ireland; local inland in southern England; abundant.
Similar species: Smaller, flatter and with sharper pronotal hind angles than *H. tardus* (18) with which it is often found. Pronotum less contracted behind and with more distinct hind angles than in the similarly-sized *H. neglectus* (11) which has the abdomen pubescent beneath.

5. *Harpalus attenuatus* Stephens

Length 7-9 mm. Black, moderately shining, females with duller elytra. Antennae entirely pale; legs brown with black femora. Sides of pronotum contracted and slightly sinuate in front of the sharp, slightly protruding hind angles; foveae (and sometimes also centre of base) punctured (Fig. 253). Elytra rather short, striae deep, seventh interval without sub-apical punctures. Wings present. On dunes and dry, sandy soils; v-viii. Local on the coast of England and Wales, inland in southern England only; scarce.
Similar species: The bicoloured legs distinguish *H. attenuatus* from nearly all mature specimens of other species. The punctured pronotal foveae resemble those of *H. rufipalpis* (14) but *H. attenuatus* has a sharper pronotal hind angle, paler antennae and no sub-apical punctures on the seventh elytral interval. Larger and with paler tibiae than *H. neglectus* (11).

6. *Harpalus dimidiatus* (Rossi)

Length 11-13.5 mm. Stout and rather convex, shining black, pronotum with a blue or green reflection basally. Antennae usually with only first segment completely pale, remainder at least partially darkened; legs black. Pronotum very wide, sides evenly rounded or somewhat narrowed in front; hind angles completely rounded, foveae shallow and often indistinct but entire basal third to quarter of pronotum densely punctured. Elytral striae deep, intervals convex apically; fifth interval with two or three, seventh interval with a long row of sub-apical punctures. Wings present. In dry, usually calcareous, grasslands and dunes; iv-viii. Very local in southern England; very scarce.
Similar species: Larger and more heavily built than all other species. The wide pronotum with curved side margins, rounded hind angles and extensive basal punctuation separate the smallest examples from the largest *H. rubripes* (13).

7. *Harpalus froelichii* Sturm

Length 8.5-10.5 mm. Male shining black, rather wide and convex; female only slightly duller. Antennae and palpi entirely pale brown; legs black with pale tarsi; front tibiae with a single row of ventral spines continuous with the pre-apical external spines; all femora with numerous ventral setae. Pronotum with sides evenly rounded and base straight; foveae linear, unpunctured; hind angles slightly rounded. Elytral striae deep, intervals slightly convex. Apical two abdominal segments with fine scattered pubescence. Wings present. On sandy lowland heaths, sand pits and dunes; vi-viii. Very local in central southern England and East Anglia; scarce, sometimes taken in light traps.

Similar species: Most resembles *H. tardus* (18) as both have a smooth pronotal base and pale antennae: distinguished by the ventral abdominal pubescence, the single row of front tibial spines and more than 10 setae (or pores) on the ventral surface of the hind femora. Larger, with paler antennae and less contracted pronotal sides than *H. neglectus* (11) which has similar abdominal pubescence and smooth pronotum. Antennae paler than *H. serripes* (15) which also sometimes has a few abdominal setae. Less elongate and with a much less heavily punctured pronotum than *H. smaragdinus* (17).

8. *Harpalus honestus* (Duftschmid)

Length 8.5-10.5 mm. Shining (male) or dull (female) green or bluish-green. Antennae darkened from second segment to apex; legs black, or with tarsi brown. Sides of pronotum narrowed and slightly sinuate in front of the sharp, rectangular hind angles; foveae punctured, otherwise base unpunctured. Elytra elongate, striae deep in male, finer in female, seventh interval with three or four sub-apical punctures. Wings absent. In sandy and chalky cliffs and grassland; v-viii. Only currently known from a probably introduced population in West Cumbria, formerly extremely local in southern and south-west England until 1905; can be abundant.

Similar species: Relatively narrower with the pronotum more contracted behind than *H. rufipalpis* (14) which is black or has at the most a slight bluish reflection. Antennae and tibiae darker than *H. rubripes* (13). See also *H. affinis* (3).

9. *Harpalus laevipes* Zetterstedt

Length 9-11.5 mm. Shining black, females only slightly duller. Appendages entirely pale. Pronotum with front margin concave, front angles protruding; sides rounded and contracted to obtuse hind angles; foveae deep and densely punctured, entire base usually somewhat roughened. Elytra almost straight-sided, intervals convex, third interval with two or three foveate punctures adjacent to second stria in apical third of elytra (Fig. 255). Wings present. On well-drained upland soils, usually in shaded sites; v-ix. Very local in northern England, local in the northern half of Scotland and recorded from just two montane localities in Ireland; scarce.

Similar species: Most resembles *H. latus* (10) but distinguished by pronotal shape, dark pronotal margins and sub-apical foveate dorsal punctures on the elytra. Less elongate and with blunter pronotal hind angles than *H. smaragdinus* (17).

10. *Harpalus latus* (Linnaeus)

Length 8.5-10.5 mm. Black or dark brown, margins of pronotum pale; females much duller than males. All appendages pale brown or yellow. Head very large relative to pronotum. Pronotum almost rectangular, front margin slightly concave, hind angles slightly rounded, basal third extensively punctured (Fig. 254). Elytra convex, sides bulging medially, intervals convex; seventh interval without sub-apical punctures. Wings present. In dry grasslands and upland heaths; v-ix. Widespread in Britain and Ireland; abundant.

Similar species: The entirely pale legs and antennae separate this species from all except *H. rubripes* (13) which has sub-apical punctures on the seventh elytral interval, and *H. smaragdinus* (17) which is more elongate, with a smaller head and pubescence on the abdominal underside. See also *H. laevipes* (9).

11. *Harpalus neglectus* Audinet-Serville

Length 6.7-8.5 mm. A rather short, convex species. Black, females very dull. Antennae pale at base, segments 2-4 or more distinctly darkened; legs black. Pronotum with front margin rather concave, front angles protruding; sides rounded, contracted to the rounded and obtuse hind angles; foveae narrow, hardly punctured. Elytral striae fine, intervals flat, without sub-apical puncture rows. Ventral abdominal segments with long, fine pubescence (Fig. 251). Wings present or absent. On sandy coastal dunes and cliffs; iv-viii. Local around the western coasts of Britain from Dorset to Cumbria; sometimes abundant.
Similar species: Smaller than *H. rufipalpis* (14) with a more constricted pronotal base, and lacking sub-apical punctures on the seventh elytral interval. See also *H. anxius* (4), *H. attenuatus* (5) and *H. froelichii* (7).

12. *Harpalus pumilus* Sturm

Length 5-6.4 mm. A short, convex, black or dark brown species. Antennae completely pale; legs and palpi brown, femora darker. Pronotal sides evenly rounded, hind angles rounded and obtuse, foveae small and unpunctured. Elytra short, rather rounded and widest medially; without pores at base of scutellary stria and on third interval (Fig. 249); apices completely rounded. Wings absent. On open, dry sandy sites; iv-vii. Extremely local in East Anglia and coastal southern England; usually scarce.
Similar species: Smaller than all other *Harpalus* species, and lacking both a scutellary pore and dorsal elytral puncture. Superficially resembles *Amara infima* but the third antennal segment is pubescent, there is only a single seta inside each eye and none at the pronotal hind angle.

13. *Harpalus rubripes* (Duftschmid)

Length 9-11.5 mm. A rather variable species; males typically shining black with a strong blue-green reflection, especially on elytra, rarely almost all black; females dull black with a hint of metallic colour on the elytra. Antennae entirely pale reddish-brown; legs vary from brown with darker femora to almost entirely black or dark brown. Pronotal sides straight or very slightly sinuate posteriorly, hind angles sharp, right-angled; base usually extensively roughened or punctured except sometimes medially. Elytra rather convex, sides slightly rounded, intervals convex, seventh interval with a row of three to seven sub-apical punctures (Fig. 250). Wings present. In open dry sandy and chalky habitats; v-ix. Widespread in England, local (and coastal) in Scotland and south-east Ireland; abundant.
Similar species: See *H. affinis* (3), *H. dimidiatus* (6), *H. honestus* (8) and *H. latus* (10).

14. *Harpalus rufipalpis* Sturm

Length 7.5-10.5 mm. Black, sometimes with a faint bluish reflection, females very dull. Antennae strongly darkened from second segment to near the apex; legs black, tarsi sometimes brown. Sides of pronotum usually slightly contracted and sinuate in front of the sharp rectangular hind angles, sometimes straighter and hind angles slightly obtuse; foveae deep and linear, pronotal base hardly punctured. Elytral intervals convex, seventh interval with a sub-apical row of two to five punctures. Wings present. In sandy places, heaths, dunes and sand pits; v-ix. Widespread but local and often coastal in England except the extreme north, Wales and the south of Ireland; scarce.
Similar species: See *H. attenuatus* (5) and *H. honestus* (8). Separated from the usually larger *H. (Cryptophonus) melancholicus* (19) by the non-pubescent tarsi and presence of sub-apical punctures on the seventh elytral interval.

15. *Harpalus serripes* (Quensel in Schönherr)

Length 8.5-11 mm. Black, rather convex. Antennae darkened from second segment to apex; legs black or with tarsi brown. Pronotum convex, sides evenly rounded, hind angles blunt and slightly obtuse; foveae narrow and shallow, rest of base unpunctured. Elytral intervals without sub-apical punctures. Wings present. On sandy or gravelly soils, usually coastal but also inland in sand and gravel pits; v-ix. Local in south Wales, southern England and East Anglia; very scarce.
Similar species: Larger than *H. tardus* (18) and with darkened second antennal segment. See also *H. froelichii* (7).

16. *Harpalus servus* (Duftschmid)

Length 7.5-8.5 mm. Wide and flat, rather dull, non-metallic. Colour varies from black with pronotal margins brown, to entirely mid brown, or with pronotal disc and head darker. Antennae pale, or slightly darkened from segments two, three or four to apex; legs ranging from entirely black to all mid brown. Pronotum very transverse, front margin angled so that anterior angles protrude; widest at or only slightly in front of the sharp, acute hind angles; hind margin evenly concave; foveae sharp and linear, base unpunctured (Fig. 257). Elytra wide and flat, striae very fine, intervals flat, without sub-apical punctures. Wings present. On dunes and sandy inland heaths; iv-viii. Very local around the coast of southern Britain from Norfolk to south Wales, also inland in Norfolk and Suffolk; very scarce.
Similar species: Superficially resembles some species of *Amara*. Flatter than other *Harpalus*, with very sharp pronotal hind angles; larger than those examples of *H. anxius* (4) that have pale antennae.

17. *Harpalus smaragdinus* (Duftschmid)

Length 9-10.5 mm. Head and pronotum very dark brown and shining, pronotal margins pale; male elytra dark and shining with a strong green metallic reflection; female elytra dull, almost black. Appendages entirely mid to pale brown. Pronotum with sides straight and slightly contracted in front of the sharp, rectangular hind angles; foveae usually wide and distinct, base distinctly punctured except sometimes medially (Fig. 252). Elytra long and parallel-sided, intervals almost flat, sub-apical punctures absent. Ventral side of abdomen with scattered pubescence. Wings present. On dry heaths, sandpits, grassland and arable fields; iv-viii. Local in England as far north as Lincolnshire, and in Wales; scarce.
Similar species: See *H. froelichii* (7), *H. laevipes* (9) and *H. latus* (10).

18. *Harpalus tardus* (Panzer)

Length 8.5-11 mm. Black and shining, female elytra dull. Antennae entirely pale; legs black with tarsi (and rarely tibial bases or femora) brown; front tibiae with a row of four to six pre-apical spines. Pronotum with sides almost straight in front of the slightly rounded hind angles; hind margin straight, base unpunctured but sometimes slightly roughened (Fig. 259). Elytral sides slightly rounded, intervals flat basally, convex apically, without sub-apical punctures. Underside of abdomen not pubescent. Wings present. In dry, open habitats including dunes, grassland, gardens and arable land; v-ix. Widespread throughout Britain except the extreme north but usually coastal in Wales, Scotland and also in the east of Ireland; abundant.
Similar species: See *H. anxius* (4), *H. froelichii* (7) and *H. serripes* (15).

3. Subgenus *Cryptophonus* Brandmayr & Zetto Brandmayr

This small group of seven European and Mediterranean species is separated from *Harpalus sensu stricto* by the pubescent tarsi (as in Fig. 247), as well as by larval characters. There are only two British species

1. Eighth elytral interval with apical row of punctures (Fig. 260) ....
   .... 19. *melancholicus* Dejean

- Eighth elytral interval without apical row of punctures ....
   .... 20. *tenebrosus* Dejean

260

19. *Harpalus melancholicus* Dejean

Length 8.5-11 mm. Black or very dark brown. Antennae brown with segments 2-4 or more darkened; legs black with tarsi paler; dorsal surface of tarsi with fine pubescence. Pronotum narrowed and convex in front, rather flat and slightly depressed inside the sharp, almost rectangular hind angles; foveae shallow, finely punctured, usually also with some punctures between these and hind angles. Eighth elytral interval with a row of two to five fine sub-apical punctures (Fig. 260). Apical abdominal segments with fine ventral pubescence. Wings present. On dunes and sandy grassland; vi-viii. Extremely local in southern England and south Wales; very scarce.
Similar species: More elongate than *H. (sensu stricto) froelichii* (7) and *H. tardus* (18). Separated from *H. tenebrosus* (20) by the sub-apical punctures on the eighth elytral interval and ventral abdominal pubescence.

20. *Harpalus tenebrosus* Dejean Plate 112

Length 8-11 mm. Black or with a slight bluish reflection, rarely with pronotal margins paler. Antennae with segments 3-6 or more darkened; legs black with tarsi paler; dorsal surface of tarsi with fine hairs. Pronotum contracted basally, sides slightly rounded, hind angles blunt and obtuse; foveae distinctly punctured, with at least some puncturing or roughening across most of the pronotal base. Elytra without sub-apical punctures on the eighth interval but rarely a single puncture is present on the fifth or seventh interval. Wings present. On dunes and sandy cliffs; v-viii. Local along the south coast of England; also occasional in south Wales and East Anglia; very scarce.
Similar species: See *H. melancholicus* (19).

2. *OPHONUS* Dejean

This Palaearctic genus of some 70 species was formerly considered as part of *Harpalus*. They are separated from the subgenus *Pseudoophonus* of *Harpalus* by having the head and pronotum punctured and pubescent as well as the elytra. Like *Harpalus* they have seed-feeding larvae. Thirteen species currently occur in Britain, with one more, *O. subsinuatus* Rey, having been found in the nineteenth century; only two are on the Irish list. Species of the subgenus *Metophonus* can be extremely difficult to identify, and the aedeagus should be checked as the most certain means of confirming a species' identity. Because of previous confusion in identifications, there has also been much duplication of names within the subgenus, some of which is still being resolved. *Ophonus (Metophonus) subsinuatus* Rey is included in the following key only, being known in Britain from three specimens from Portland, Dorset in 1886.

Key to species of *Ophonus*

1. Hind angles of pronotum rounded or bluntly angled, sides not (or hardly) sinuate (Plate 113, Figs 261, 266) 2

- Hind angles of pronotum sharp, right-angled or obtuse but distinct, sides more sinuate or contracted basally in front of hind angles (Plate 114, Figs 269, 271, 273) 5

2. Length less than 9 mm; sides of pronotum slightly sinuate in front of distinct but blunt hind angles (Fig. 261) ....2. *azureus* (F.)

- Length 10 mm or more; sides of pronotum completely rounded or straight in front of rounded hind angles (Fig. 266) 3

3. Elytral apices blunt, more sinuate sub-apically (Fig. 262); entire body usually with a green reflection; elytral pubescence dark; aedeagal apex long, four times as long as wide in dorsal view (Fig. 263) .... 4. *stictus* Stephens

- Elytral apices sharp, less sinuate in front of this (Fig. 264); elytra usually with a blue reflection, head and pronotum black; elytral pubescence pale; aedeagal apex shorter 4

4. Sides of pronotum evenly rounded, hind angles hardly evident (Plate 113); length usually less than 12 mm; aedeagal apex short, twice as long as wide in dorsal view (Fig. 265) ....
.... 1. *ardosiacus* (Lutschnik)

- Sides of pronotum straight and contracted in front of more evident obtuse, blunt hind angles (Fig. 266); length usually more than 12 mm; aedeagal apex intermediate, three times as long as wide in dorsal view (Fig. 267) .... 3. *sabulicola* (Panzer)

5. With a distinct metallic, usually green, reflection ....
.... 6. *laticollis* Mannerheim

- Entirely non-metallic, from reddish-brown to black 6

6. Elytral stria finely punctured; aedeagus with straight, downwardly deflected apex without apical disc (Fig. 268) ....
.... 5. *cordatus* (Duftschmid)

- Elytral striae not punctured 7

7. Elytral shoulders rounded, without any protruding tooth; pronotal sides strongly contracted towards base, with slight sinuation in front of hind angles (Fig. 269); aedeagal apex broad with apical disc protruding both above and below (Fig. 270) .... 12. *rupicola* (Sturm)

- Elytral shoulders angled, usually with a protruding tooth (Fig. 271); pronotal sides less contracted basally; aedeagus not as above 8

8. Pronotum narrow, less than 1.3 times as wide as long, sides little contracted or sinuate towards base (Fig. 271); aedeagus very slender with apical disc reduced to slight lateral projections, not protruding dorsally or ventrally (Fig. 272) ... 9. *puncticeps* Stephens

- Pronotum wider, more than 1.3 times as wide as long, and more sinuate laterally; aedeagus either with apical disc protruding dorsally and/or ventrally (Figs 278, 281, 282) or smooth without apical disc (Figs 275, 277) 9

9. Base of pronotum without any trace of a raised border or rim (Fig. 273) 10

- Base of pronotum with a fine border or at least a trace of one laterally, sometimes missing medially (Figs 274, 280) 13

10. Side of pronotum with two long setae (Fig. 273); apex of aedeagus smooth and deflected downwards, without well-defined apical plate or disc (Figs 275, 277) 11

- Side of pronotum with a single long seta (Fig. 274); aedeagus with an apical disc, or at least expanded apically, not deflected downwards (Figs 278, 281, 282) 12

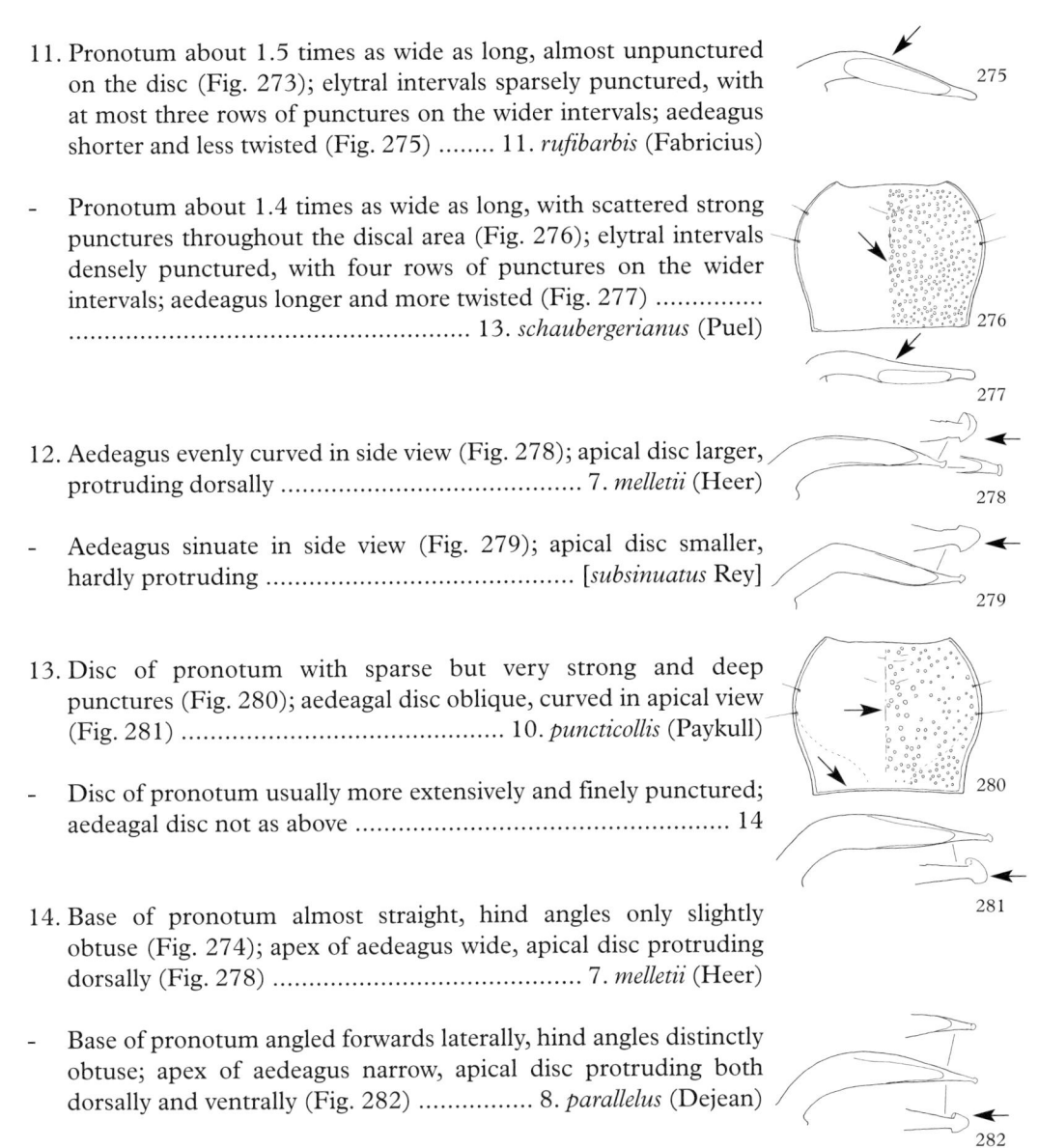

11. Pronotum about 1.5 times as wide as long, almost unpunctured on the disc (Fig. 273); elytral intervals sparsely punctured, with at most three rows of punctures on the wider intervals; aedeagus shorter and less twisted (Fig. 275) .... 11. *rufibarbis* (Fabricius)

- Pronotum about 1.4 times as wide as long, with scattered strong punctures throughout the discal area (Fig. 276); elytral intervals densely punctured, with four rows of punctures on the wider intervals; aedeagus longer and more twisted (Fig. 277) .... 13. *schaubergerianus* (Puel)

12. Aedeagus evenly curved in side view (Fig. 278); apical disc larger, protruding dorsally .... 7. *melletii* (Heer)

- Aedeagus sinuate in side view (Fig. 279); apical disc smaller, hardly protruding .... [*subsinuatus* Rey]

13. Disc of pronotum with sparse but very strong and deep punctures (Fig. 280); aedeagal disc oblique, curved in apical view (Fig. 281) .... 10. *puncticollis* (Paykull)

- Disc of pronotum usually more extensively and finely punctured; aedeagal disc not as above 14

14. Base of pronotum almost straight, hind angles only slightly obtuse (Fig. 274); apex of aedeagus wide, apical disc protruding dorsally (Fig. 278) .... 7. *melletii* (Heer)

- Base of pronotum angled forwards laterally, hind angles distinctly obtuse; apex of aedeagus narrow, apical disc protruding both dorsally and ventrally (Fig. 282) .... 8. *parallelus* (Dejean)

Subgenus *Ophonus sensu stricto*

1. *Ophonus ardosiacus* (Lutschnik) Plate 113

Length 9.5-11.5 mm. Head and pronotum black, elytra with a distinct dark blue metallic reflection. Appendages uniformly reddish-brown. Pronotum with sides almost evenly rounded, only slightly narrower basally, hind angles very rounded and hardly distinct; punctures on disc smaller and more scattered than near the margins. Elytra long and parallel-sided, striae unpunctured, intervals densely punctured, pubescence pale; apices sharp, the margins in front of this hardly sinuate (Fig. 264). Aedeagal apex relatively short, about twice as long as wide, with small, ventrally protruding, apical disc (Fig. 265). Wings present. In open habitats on chalk or limestone soils, also in coastal clay cliffs; v-viii. Local in England and south Wales as far north as the Humber; sometimes abundant. **Similar species:** Smaller than *O. sabulicola* (3) with sides of the pronotum more evenly rounded and the hind angles less evident. Separated from *O. stictus* (4) by the smaller size and blue metallic colour, the pale elytral pubescence and the sharp elytral apices.

2. *Ophonus azureus* (Fabricius)

Length 7-8.5 mm. Black with a distinct green reflection, especially on elytra; rarely dark reddish-brown, non-metallic (var. *similis* Dejean). Appendages light brown. Sides of pronotum slightly contracted and just sinuate in front of blunt, obtuse hind angles (Fig. 261); disc with deep, well-scattered punctures; base with distinct raised border. Punctures on elytral intervals sometimes incorporated into the striae, which appear partly punctured; pubescence pale, apices almost sharp, sub-apical margin hardly sinuate. Aedeagus with well-developed, oblique apical disc. Wings present or absent. In open coastal sites, also inland on warm chalk or limestone slopes; v-viii. Local in southern England and on the Welsh coast; scarce.
Similar species: The metallic green colour distinguishes *O. azureus* from all other small *Ophonus*. Rare non-metallic individuals resemble *O. parallelus* (8) and *O. puncticollis* (10) in size and punctuation but have the pronotal hind angles bluntly rounded. Usually smaller than the similarly coloured *O. laticollis* (6), which has sinuate pronotal sides.

3. *Ophonus sabulicola* (Panzer)

Length 12-17 mm. Head and pronotum black or dark brown, elytra dark with a distinct metallic blue reflection. Appendages mid brown. Pronotum large and wide, widest well in front of middle and strongly contracted behind to the very obtuse and blunt but distinct hind angles (Fig. 266). Elytral punctuation moderate, pubescence pale; apices sharp, not divergent, sub-apical marginal sinuation almost absent. Aedeagal apex moderately elongate, about three times as long as wide (Fig. 267). Wings present. On dry chalky or sandy soils, usually coastal; vii-ix. Very local in southern England and south Wales; scarce.
Similar species: The blue rather than greenish elytra distinguish it from the similarly sized *O. stictus* (4) from which it also differs by features of the pronotal hind angles and elytral apices. See also *O. ardosiacus* (1).

4. *Ophonus stictus* Stephens

Length 12-16 mm. Black or dark brown with a distinct greenish metallic reflection. Appendages mid to pale brown. Pronotum widest in front of middle, sides contracted behind to the blunt but distinct hind angles. Elytra long, slightly widened medially, finely punctured with dark brown pubescence; apices slightly rounded and diverging, and with a distinct sub-apical sinuation (Fig. 262). Aedeagal apex very elongate, at least four times as long as wide (Fig. 263). Wings present. In quarries and on limestone or chalk soils; vi-viii. Extremely local in southern and eastern England; very scarce.
Similar species: See *O. ardosiacus* (1) and *O. sabulicola* (3).

Subgenus *Metophonus* Bedel

5. *Ophonus cordatus* (Duftschmid)

Length 7.5-10 mm. Reddish to dark brown; appendages mid to light brown. Pronotum 1.3 times as wide as long, strongly contracted towards base, sides slightly sinuate just in front of the rectangular hind angles; base bordered, hind angles level with the interval between the fifth and sixth elytral striae. Elytral shoulders rounded, without protruding shoulder teeth; striae finely but distinctly punctured, wider intervals with three or four irregular rows of punctures. Aedeagus without apical disc (Fig. 268). Wings present. In dunes and coastal cliffs; viiii-x. Restricted to the south coast of England from Dorset to Kent; extremely scarce.

Similar species: Most resembles *O. rupicola* (12) as both have rounded elytral shoulders without protruding teeth, and strongly contracted pronotal sides. But *O. cordatus* has punctured striae, its pronotal sides are less sinuate and the pronotal base is clearly bordered. Pronotum more constricted basally than *O. puncticeps* (9) which does not have punctured striae. The aedeagus resembles those of *O. rufibarbis* (11) and *O. schaubergerianus* (13) in having no apical disc but is intermediate in length between those two species.

6. *Ophonus laticollis* Mannerheim

Length 8-11 mm. Black with a distinct green metallic reflection. Appendages brown. Pronotum about 1.4 times as wide as long, sides slightly to moderately sinuate in front of sharp, rectangular hind angles; base straight, without basal border; disc rather sparsely punctured. Elytra with distinct shoulder tooth, sides slightly rounded, widest at middle, intervals densely punctured; striae not punctured. Wings present. On dry, usually chalky soils, with some vegetation; v-ix. Extremely local in south and east England, and in south Wales; very scarce.

Similar species: Separable from all other species of subgenus *Metophonus* by the metallic green dorsal surface. See also *O. azureus* (2).

7. *Ophonus melletii* (Heer)

Length 6-8 mm. Mid to dark brown, elytra usually darker than head and pronotum. Appendages pale brown. Pronotum 1.3-1.4 times as wide as long (Fig. 274), sides moderately sinuate, with one long and two or three small lateral setae in anterior half; disc finely but evenly punctured; base straight, hind angles only slightly obtuse; basal border usually present at least laterally but sometimes very indistinct. Elytral shoulders toothed; striae impunctate, intervals with three or four irregular rows of punctures. Aedeagal apex wide; apical disc protruding dorsally but hardly at all ventrally, not curved in apical view (Fig. 278). Wings present. In open or semi-shaded habitats on chalky or sandy soils often on the coast; v-viii. Very local but widespread in England south of a line from the Humber to the Severn; usually scarce.

Similar species: On average slightly larger and paler than *O. parallelus* (8), with straighter hind pronotal margin and more setae on the sides of the pronotum. Pronotal disc less strongly punctured than in *O. puncticollis* (10). Specimens with the basal border of the pronotum reduced can be distinguished from *O. rufibarbis* (11) and *O. schaubergerianus* (13) by their smaller size, less transverse pronotum with a single large lateral seta and the distinct aedeagal disc.

8. *Ophonus parallelus* (Dejean)

Length 5.2-7.5 mm. Uniformly mid brown to almost black; appendages mid brown. Pronotum 1.4 times as wide as long, sides slightly sinuate, with one or two lateral setae in anterior half; disc with well-scattered fine punctures; base finely bordered, hind margin angled forwards laterally, so that hind angles are clearly obtuse and slightly rounded. Elytral shoulder teeth distinct; striae impunctate, wider intervals with four irregular rows of punctures. Aedeagal apex narrow, apical disc protruding both dorsally and ventrally, not curved in apical view (Fig. 282). Wings present. On open, usually chalky soils, mainly on the coast; v-viii. Extremely local in south-east England; very scarce.

Similar species: See also *O. azureus* (2) and *O. melletii* (7). Pronotal disc less strongly punctured than in *O. puncticollis* (10). Smaller than *O. rufibarbis* (11) with less transverse pronotum that has a fine but evident basal border.

9. *Ophonus puncticeps* Stephens Plate 114

Length 6.5-9 mm. Uniformly mid to dark brown; appendages light yellowish brown. Pronotum narrow, less than 1.3 times as wide as long (Fig. 271), sides only slightly sinuate, with a single long lateral seta; hind angles level with interval between fifth and sixth elytral striae; basal margin with fine but distinct border. Elytra long and parallel-sided, shoulders minutely toothed; wider intervals with three or four irregular rows of punctures. Aedeagus with slender apex which is slightly serrated dorsally, apical disc reduced to a slight lateral and downward protrusion (Fig. 272). Wings present. On open, dry, usually sandy soils, including cultivated fields; v-ix. Widespread in southern England, more local and usually coastal in Wales, south-east Ireland and further north in England; sometimes abundant.
Similar species: More slender and parallel-sided than *O. rufibarbis* (11) and *O. schaubergerianus* (13) with a distinct hind border on the less transverse pronotum. Pronotum less constricted basally than *O. rupicola* (12). See also *O. cordatus* (5).

10. *Ophonus puncticollis* (Paykull)

Length 6.5-9 mm. Mid brown to nearly black; appendages mid to dark brown. Pronotum rather variable, 1.3-1.4 times as wide as long, sometimes very wide, with convex disc and sides variably sinuate in front of sharp hind angles; base finely bordered, sides with at least two long setae; disc with exceptionally strong but widely scattered punctures (Fig. 280). Elytral shoulder tooth distinct, sides rounded, intervals densely punctured. Aedeagus with long, slender apex; apical disc small but distinct, extending obliquely both dorsally and ventrally, curved in apical view (Fig. 281). Wings present. On dry chalk, limestone or gravelly soils, often semi-shaded; v-viii. Extremely local in south and east England and on the coast of south Wales; very scarce.
Similar species: See *O. azureus* (2), *O. melletii* (7) and *O. parallelus* (8).

11. *Ophonus rufibarbis* (Fabricius)

Length 6.5-9.5 mm. Mid to dark brown, elytra darker than forebody in paler examples. Appendages yellow or pale brown. Pronotum about 1.5 times as wide as long, sides strongly sinuate, hind angles sharp; sides with two long setae; base without border; disc very sparsely punctured with fine punctures (Fig. 273). Elytra with shoulder teeth, sides slightly rounded, intervals finely punctured with about three irregular rows of punctures on the wider ones. Aedeagus only moderately long, almost straight in dorsal view, without apical disc, apex bent ventrally (Fig. 275). Wings present. In partly vegetated dry habitats on almost all soils; iv-viii. Widespread in England, more local in the rest of Britain (except the extreme north) and the east of Ireland; abundant.
Similar species: Separated from *O. schaubergerianus* (13), which also has a non-bordered pronotal base, by the more transverse pronotum, and sparser punctuation on the pronotal disc and elytral intervals. The aedeagus resembles that of *O. schaubergerianus* in having no apical disc but is shorter and straighter viewed from above. See also *O. cordatus* (5), *O. melletii* (7), *O. parallelus* (8) and *O. puncticeps* (9).

12. *Ophonus rupicola* (Sturm)

Length 7-10 mm. Mid brown to almost black, elytra with a slight metallic reflection; appendages brown. Pronotum less than 1.3 times as wide as long, sides strongly contracted towards base but sinuate in front of the sharp hind angles (Fig. 269); base not bordered, hind angles level with, or just outside, the fifth elytral stria. Elytral shoulders rounded, without teeth; striae unpunctured, wider intervals with three irregular rows of punctures. Aedeagus with wide, curved apical disc that extends slightly both dorsally and ventrally (Fig. 270). Wings present. On dry, open usually calcareous ground inland, as well as coastal habitats; iv-viii. Local in southern England and East Anglia; scarce.

Similar species: Distinguished from other species without a basal pronotal border, *O. rufibarbis* (11) and *O. schaubergerianus* (13), by the rounded elytral shoulders without protruding teeth, the more strongly contracted pronotal sides and the aedeagus with an apical disc. See also *O. cordatus* (5) and *O. puncticeps* (9).

13. *Ophonus schaubergerianus* (Puel)

Length 8-10 mm. Mid to dark reddish brown, with elytra usually darker than the forebody; appendages brown. Pronotum 1.4 times as wide as long (Fig. 276), sides moderately sinuate and with two long setae, base not bordered; disc with moderately dense strong punctures. Elytra long, sides slightly rounded, shoulder tooth present; intervals densely punctured, wider intervals with four or even five irregular rows of punctures. Aedeagus very long, sinuate in dorsal view, without apical disc but apex bent downwards (Fig. 277). Wings present. In dry grasslands and open scrub or woodland; v-viii. Local in south and east England as far north as Yorkshire, one coastal record from Wales; usually scarce.

Similar species: See *O. cordatus* (5), *O. melletii* (7), *O. puncticeps* (9) and *O. rufibarbis* (11). On average larger and with more densely punctured elytral intervals than other species of the subgenus.

3. *ANISODACTYLUS* Dejean

This Holarctic genus has more than 50 species, of which ten occur in Europe but only three in Britain and two in Ireland. Separated from *Harpalus* by the shorter hind tibial spur (Fig. 245) and the presence of dense adhesive setae underneath the front tibiae of males. Also usually with a small, diffuse pale reddish mark on the head mid-way between the eyes.

Key to species of *Anisodactylus*

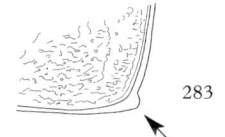

1. Upper surface metallic golden-green; hind angles of pronotum rounded .... 3. *poeciloides* (Stephens)

- Upper surface black; hind angles of pronotum sharp with a small protruding tooth (Fig. 283) 2

2. Length usually more than 10 mm; entire apical region of elytra finely pubescent (Plate 115) .... 1. *binotatus* (Fabricius)

- Length usually less than 10 mm; only outer two or three elytral intervals pubescent .... 2. *nemorivagus* (Duftschmid)

1. *Anisodactylus binotatus* (Fabricius) Plate 115

Length 10-13 mm. Dorsal surface black except for the pale spot on the head; head and pronotum shining, elytra duller. Antennae usually black with segments 1-2 pale but occasionally entirely pale; legs typically black, rarely pale brown (var. *spurcaticornis* Dejean). Head unpunctured, or with extremely fine punctures. Pronotum transverse, sides with a single long lateral seta, hind angles with a small tooth (Fig. 283); entire base densely and roughly punctured; anterior third also often with fine punctures. Elytral shoulders rounded; striae unpunctured, intervals convex, with scattered punctures at the base of each interval; outer two intervals and apical region of elytra with fine golden pubescence; apices diverging, with oblique sub-apical sinuation. Wings present. In damp meadows and marshy habitats, as well as arable land on poorly-draining soils; iv-viii. Widespread in England and Wales and the west of Ireland, very local in lowland Scotland and the east of Ireland; sometimes abundant.
Similar species: Larger than *A. nemorivagus* (2) with more widespread elytral pubescence apically. Most individuals have black legs, whereas in *A. nemorivagus* these are typically pale but both species have varieties with the 'opposite' leg colour. The shape, colour and pronotal features of *A. binotatus* superficially resemble some *Pterostichus* such as *P. melanarius* but it can be recognised by the pubescent third antennal segment.

2. *Anisodactylus nemorivagus* (Duftschmid)

Length 8-10 mm. Black with usually a very faint pale mark on the head. Antennae black with first segment red; legs usually reddish-brown or rarely black (var. *atricornis* (Stephens)). Pronotal sides strongly contracted to the obtuse but sharp hind angles which are hardly toothed. Elytral shoulders angled between basal rim and elytral margin; intervals without punctures at their base; outer two elytral intervals finely pubescent, especially towards the rear; apices not or little divergent, sub-apical sinuation strong. Wings present. On dry sandy heaths; iv-vii. Very local in southern England and south Wales, with a single Irish record from Cork; scarce.
Similar species: See *A. binotatus* (1).

3. *Anisodactylus poeciloides* (Stephens)

Length 9-12.5 mm. Shining metallic coppery green. Appendages black except for pale first antennal segment; apical spur of front tibiae three-toothed. Pronotal sides rounded, hind angles curved and hardly distinct; base of pronotum with coarse rugose punctuation, sides, anterior region and part of disc sometimes also finely punctured. Elytral shoulders bluntly angled, bases of intervals with scattered punctures, outer two intervals with dense, semi-erect pale pubescence; apices rounded, slightly diverging, with rather transverse sub-apical sinuation. Wings present. Under stones and litter on open, saline coastal habitats including the upper parts of saltmarshes; iv-viii. Very local around the south and east coasts of England from Dorset to Suffolk; usually scarce.
Similar species: Superficially resembles *Harpalus affinis*, from which it can be distinguished by the three-toothed tibial spur. Species of *Poecilus* are similar but have non-pubescent third antennal segments, sharp pronotal hind angles and two setiferous punctures inside each eye.

4. *DIACHROMUS* Erichson

This characteristically coloured genus has a single species widely distributed throughout Europe (except the north) and the Middle East. Doubtfully established in Britain.

1. *Diachromus germanus* (Linnaeus)

Length 7.5-10 mm. Head and front half of elytra reddish brown; pronotum, scutellum and apical mark on elytra dark, usually metallic greenish-blue. Appendages pale brown with apical antennal segments and front tibial spur darker. Entire dorsal surface punctured and with long pale pubescence. Pronotum strongly contracted behind, sides sinuate just in front of sharp hind angles; hind margin angled forwards laterally. Elytral shoulders rounded, apical margin rather transversely oblique and slightly sinuate in front of sharp apices. Occasional solitary specimens found in dry habitats in southern England from Kent to Cornwall; presumed extinct until a recent occurrence in Sussex.

5. *SCYBALICUS* Schaum

There is a single species occurring in western Europe and the Mediterranean region. The pubescent upper surface resembles that of a large *Ophonus* but the antennal segments are all densely pubescent, and the base of each elytron has a characteristic depression and sinuate border (Fig. 242).

1. *Scybalicus oblongiusculus* (Dejean)

Length 11-13 mm. Dorsal surface and appendages uniformly mid to dark brown; head and pronotum densely and rather coarsely punctured, elytra finely punctured and pubescent on all intervals. All antennal segments and tarsi finely pubescent. Pronotum widest in front of middle and very strongly contracted to the obtuse hind angles; sides with a single long seta, hind angles without setae. Elytra elongate and parallel-sided; base of each elytron depressed between scutellum and elytral shoulder, with the basal margin sinuate in front of third stria. Wings present. Typically in dry, open situations but the breeding habitat is not known in Britain. Dorset coast, last seen in 1951; more recent occurrences in Surrey and Essex.

6. *DICHEIROTRICHUS* Jacquelin du Val

The species of this Palaearctic genus of about 12 species (five in Europe) are usually restricted to saline habitats. The two British species are distinguished by the entirely pubescent upper surface and lack of any scutellary stria. Only one occurs in Ireland.

Key to species of *Dicheirotrichus*

1. Elytra less densely punctured, interval 1 with a single row, wider intervals with two rows, of punctures (Plate 116) .... 1. *gustavii* Crotch

- Elytra more densely punctured, interval 1 with a double row, wider intervals with three rows of punctures .... 2. *obsoletus* (Dejean)

1. *Dicheirotrichus gustavii* Crotch Plate 116

Length 5.5-7.5 mm. Colour variable, from entirely pale yellow-brown, to almost entirely black with pale margins on head, pronotum and elytral base; males usually darker than females. Appendages yellow, femora sometimes darkened basally in males. Pronotal sides strongly contracted and slightly sinuate basally, hind angles sharp, slightly protruding. Elytral pubescence long and semi-erect, adjacent hairs overlapping; inner intervals with a single row of punctures and hairs, wider intervals with a double row at most. Wings present. In saltmarshes and under shore debris; vi-ix. Widespread around the coasts of Britain and Ireland; very abundant.
Similar species: On average smaller and elytra narrower than in *D. obsoletus* (2), more variable in colour and less densely punctured.

2. *Dicheirotrichus obsoletus* (Dejean)

Length 6-8 mm. Pale yellow-brown, with at most a darkened elongate streak on each elytron; appendages pale. Pronotal sides only moderately contracted basally but sinuate in front of the protruding hind angles. Elytral sides rounded, pubescence hardly overlapping; inner intervals with a double row of punctures and hairs, wider intervals with three rows at least. Wings present. In saltmarshes; v-viii. Local around English south and east coasts from Cornwall to Lincolnshire; sometimes abundant.
Similar species: See *D. gustavii* (1).

7. TRICHOCELLUS Ganglbauer

Sometimes included as a sub-genus of *Dicheirotrichus* but the two British and Irish species are much smaller, with less extensive elytral pubescence. As in that genus, there is no scutellary stria but in contrast most of the upper surface is smooth and unpunctured. There are about 30 species, mostly Palaearctic.

Key to species of *Trichocellus*

1. Second antennal segment black, contrasting strongly with the pale first segment .... 1. *cognatus* (Gyllenhal)

- Second antennal segment pale or mid brown, similar to the first segment .... 2. *placidus* (Gyllenhal)

1. *Trichocellus cognatus* (Gyllenhal)

Length 4-5.5 mm. Typically with head and pronotum (except paler margins) black; elytra brown with a variable-sized black patch extending over much of the apical two-thirds, except marginally and along the sutural interval. Antennae black with first segment clear reddish brown; legs mid to dark brown or black with paler tibiae, at least basally. Pronotum moderately widened in front, sides usually slightly rounded in front of very obtuse and indistinct hind angles. Elytra with outer two intervals and apical region distinctly pubescent. Wings present. On moors and heaths, usually with *Calluna*; iii-v, viii-x. Widespread in Britain north of a line from the Humber to the Severn, local in Wales and Ireland; often abundant.
Similar species: Usually with more extensive dark markings than *T. placidus* (2) and with more distinct elytral pubescence at the sides. The pronotum of *T. cognatus* is slightly less widened in front but all these features vary in both species and are comparative.

2. *Trichocellus placidus* (Gyllenhal) Plate 117

Length 4-5.5 mm. Typically head and pronotum brown with a dark central mark of varying size; elytra mid to light brown with an elongate darker patch on the sub-apical part of intervals 2-4. Antennae pale brown or slightly darkened from fourth segment to apex; legs uniformly pale to mid brown. Pronotum rather strongly widened in front, sides almost straight in front of the obtuse hind angles. Pubescence on elytral margins and apex fine and rather sparse. Wings present. In well-vegetated lowland marshes, damp grasslands and moist woodland litter; iii-v, viii-x. Widespread in Britain and Ireland, especially in the east; abundant.

Similar species: See *T. cognatus* (1).

8. *BRADYCELLUS* Erichson

Small, rather convex harpalines. They are characterised by the toothed mentum (Fig. 246), pale antennae and non-pubescent elytra with a scutellary stria. There are more than 100 species worldwide. Some species, especially *B. harpalinus*, are rather variable; the difficulty of their identification has led to some duplication of specific names. Five of the seven British species occur in Ireland.

Key to species of *Bradycellus*

1. Hind angles of pronotum completely rounded and not distinct, sides rounded or straight in hinder half, not at all sinuate (Plate 118, Figs 284, 286) 2

- Hind angles of pronotum distinct, even if not sharp, sides in front of these at least slightly sinuate (Figs 289, 290) 4

2. Scutellary stria longer and usually well-developed, twice as long as width of second elytral interval (Fig. 284); eyes large and more protruding; length usually more than 4 mm; aedeagus slender, with internal teeth (Fig. 285) .... 4. *harpalinus* (Audinet-Serville)

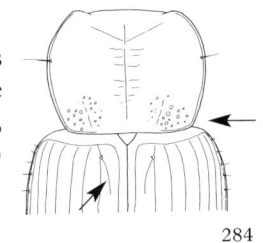

- Scutellary stria usually short, interrupted or almost absent, hardly longer than width of second elytral interval (Fig. 286); eyes smaller and flatter; length often less than 4 mm; aedeagus not as above .....
.... 3

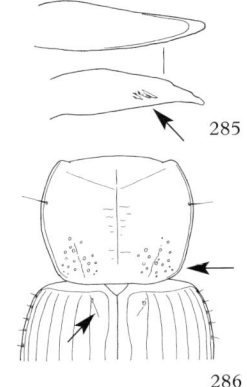

3. Colour pale reddish-brown; pronotum more contracted towards base (Fig. 286); wings usually absent; aedeagus slender, with internal fold (Fig. 287) .... 1. *caucasicus* (Chaudoir)

- Colour black or very dark brown, elytral suture paler; pronotum less contracted towards base; wings probably usually present; aedeagus broader, without internal teeth or fold (Fig. 288) .... 2. *csikii* Laczó

4. Length less than 3.5 mm .... 5. *ruficollis* (Stephens)

- Length more than 4 mm 5

5. Entire basal third and median anterior region of pronotum with extensive punctures (Fig. 289) .... 3. *distinctus* (Dejean)

- Base of pronotum less punctured, especially centrally; front of pronotum with at most a few scattered punctures 6

6. Colour pale reddish-brown; elytral sides more or less parallel (Fig. 290); wings present .... 7. *verbasci* (Duftschmid)

- Colour dark reddish brown to almost black; sides of elytra rounded (Fig. 291); wings usually absent .... 6. *sharpi* Joy

1. *Bradycellus caucasicus* (Chaudoir)

Length 3.5-4 mm. Pale to mid reddish-brown, elytral suture sometimes paler. Appendages pale. Eyes only moderately protruding. Pronotum with sides straight or slightly rounded behind, strongly contracted towards base; side borders not reaching around the hind angles as far as basal foveae (Fig. 286). Elytra rather short and narrow in proportion to the pronotum; scutellary striae oblique, short or even partially lacking; striae very finely punctured in places but intervals without micro-punctures. Aedeagal apex sharp and symmetrical in dorsal view; internal sac with a convex fold (Fig. 287). Wings usually absent. On dry heaths usually with *Calluna* on sand or gravel soils; iii-iv, vii-x. Widespread in northern Britain, local in the south and in Ireland; scarce.
Similar species: Much paler in colour than *B. csikii* (2) with the pronotum more contracted basally, and relatively shorter elytra. Overlaps in size with small examples of *B. harpalinus* (4) but these are usually winged, with a longer scutellary stria, and proportionally longer and wider elytra compared to the pronotum, as well as having larger eyes. Paler and with the pronotal sides less sinuate than *B. ruficollis* (5).

2. *Bradycellus csikii* Laczó

Length 3.5-4.5 mm. Black or very dark brown, elytral margins and suture paler. Appendages pale. Pronotal sides straight and only slightly contracted to the rounded hind angles, hind margin as wide as front margin; with scattered large punctures in the foveae but not medially; side borders reaching around the hind angles as far as basal foveae. Elytra clearly wider than pronotum, intervals with scattered micro-punctures. Aedeagal apex wide, internal sac without teeth (Fig. 288). Wings usually present. There is an old record from Surrey, and recent records from sandy soils in Suffolk; very scarce.
Similar species: See *B. caucasicus* (1). Smaller and darker on average than *B. harpalinus* (4) which lacks the micro-punctures on the elytral intervals and has a longer scutellary stria.

3. *Bradycellus distinctus* (Dejean)

Length 4-5 mm. Mid reddish-brown, rarely lighter or darker; appendages pale. Pronotum sinuate in front of sharp hind angles; with extensive and strong punctures in the front depression and across the basal third including the foveae (Fig. 289). Elytra rather convex and rounded at sides, scutellary stria long, sometimes reduced on one side; third interval without dorsal puncture. Wings usually absent. On sandy and gravelly sites usually near the coast; iii-v, viii-x. Once very local around the English south coast from Cornwall to Norfolk, also Lancashire; now only at Dungeness, Kent; very scarce.
Similar species: The sinuate pronotal sides and sharp hind angles resemble those of *B. sharpi* (6) and *B. verbasci* (7) but *B. distinctus* is the only species to have extensive punctuation across both the front margin and the basal third of the pronotum. It is also the only species without any dorsal puncture on the third elytral interval.

4. *Bradycellus harpalinus* (Audinet-Serville) Plate 118

Length 3.8-5 mm. A rather variable species; typically mid reddish-brown, with pronotum, elytral margins and suture slightly paler than remainder of upper surface; extreme forms range from entirely pale brown to almost black. Appendages pale, or rarely antennae slightly darkened from the fourth segment to apex. Eyes large and protruberant. Pronotum with sides contracted and slightly rounded in front of rounded hind angles; side borders extending along the base to at least level with the middle of the punctured foveae (Fig. 284). Elytra with long scutellary stria, almost parallel to stria 2; third interval with a dorsal puncture in apical third; striae unpunctured and intervals without micro-punctures. Aedeagal apex sharp but not symmetrical in dorsal view; internal sac with slender teeth apically (Fig. 285). Wings present. In a wide range of dry habitats including gardens, grasslands, heath, arable land and woodlands; vii-x. Widely distributed throughout Britain and Ireland; abundant.
Similar species: See *B. caucasicus* (1) and *B. csikii* (2). Separated from the other common species *B. sharpi* (6) and *B. verbasci* (7) by the non-sinuate pronotal sides and rounded hind angles.

5. *Bradycellus ruficollis* (Stephens)

Length 2.5-3.4 mm. Dark brown to almost black, pronotum and especially elytral suture paler. Appendages pale. Eyes large but rather flat. Pronotum strongly contracted behind, sides slightly sinuate in front of the very obtuse but distinct hind angles; base hardly punctured. Scutellary stria rather weak but moderately long; elytra slightly sinuate before apices. Wings usually absent. On well-draining heaths and moors with *Calluna*; iii-v, viii-x. Widespread in the north of Britain and the east of Ireland, more local in Wales, the south of England and western Ireland; often abundant.

Similar species: Smaller than all other species of the genus, and the only one in which the males have both the middle and front tarsi expanded. See also *B. caucasicus* (1).

6. *Bradycellus sharpi* Joy

Length 4-4.8 mm. Dark red brown to almost black, elytra usually darker than head and pronotum. Appendages pale. Pronotal sides distinctly sinuate in front of the obtuse but sharp hind angles; side borders hardly extending onto hind margin, which is angled forwards only slightly on each side; basal punctuation variable but always present in the foveae. Elytra rather globular, sides rounded (Fig. 291); scutellary stria medium-sized but rather oblique. Wings absent or present. In damp grasslands, woodland litter and other damp, shaded sites; iii-v, viii-x. Widespread in Britain and Ireland, mainly near the coast in Wales and Scotland; abundant.
Similar species: See *B. distinctus* (3) and *B. harpalinus* (4). Darker in colour than *B. verbasci* (7) which has more parallel-sided elytra.

7. *Bradycellus verbasci* (Duftschmid)

Length 4-5 mm. Pale red brown, elytra usually darker apically. Appendages pale. Pronotal sides distinctly sinuate in front of the obtuse but sharp hind angles; side borders extending onto hind margin, which is sharply angled forwards on each side; basal punctuation often only present in the foveae. Elytra with sides almost straight (Fig. 290), scutellary stria long and nearly parallel to stria 2. Wings present. On open, usually well-drained soils, often on arable fields and waste ground; iii-iv, viii-x. Widespread in England, local in Wales, Scotland and Ireland; abundant.
Similar species: See *B. distinctus* (3), *B. harpalinus* (4) and *B. sharpi* (6).

9. *STENOLOPHUS* Dejean

A world-wide but mainly Holarctic genus of about 70 species, of which only three occur in Britain and one in Ireland. Records of *S. plagiatus* Gorham were of the introduced north American *Agonoderus comma* (Fabricius), which has not become established here. *Stenolophus* species are distinguished from *Bradycellus* and *Acupalpus* by their larger size and iridescent microsculpture on the elytra. The presence of a scutellary stria and lack of elytral pubescence separate them from *Dicheirotrichus* and *Trichocellus*. All are winged and can fly.

Key to species of *Stenolophus*

1. Pronotum black with brown margins (Plate 119) .... 1. *mixtus* (Herbst)

- Pronotum mainly reddish-brown, rarely darkened medially or along front margin 2

2. Apical two-thirds of elytra with a well-delimited dark mark extending almost to the outer interval; antennae black with extreme apex and basal two segments pale .... 3. *teutonus* (Schrank)

- Elytra entirely pale, or with a diffuse darker area on apical third; antennae mostly brown, only the median segments dark brown, rarely black .... 2. *skrimshiranus* Stephens

1. *Stenolophus mixtus* (Herbst) Plate 119

Length 5-6 mm. Head black, pronotum black with brown margins, elytra dark brown, usually with margins, suture and an area around the shoulder paler. Antennae black or dark brown with first segment pale; legs entirely pale yellow-brown. Pronotum only moderately transverse, sides slightly rounded, base rugosely punctured, except sometimes medially. Elytra moderately iridescent, striae unpunctured, intervals concave. In marshes and at the edges of standing water, especially on clay soils; iv-vii. Widely distributed in most of England, except the extreme north and west; local in Wales and Ireland; abundant.
Similar species: The darkened pronotum distinguishes this species from the other two species of the genus.

2. *Stenolophus skrimshiranus* Stephens

Length 5.5-6.5 mm. Head black, pronotum and elytra reddish brown, often with a diffuse darker area towards elytral apices. Antennae with basal two segments pale, remainder darkened especially medially but apical segments usually mid brown; legs yellow or pale brown. Pronotum very transverse, sides almost evenly rounded, side borders extending just on to hind margin; basal region with scattered punctures except in the foveae. Elytra strongly iridescent, intervals flat. In marshes and near standing water where there is vegetation and litter; iv-vii. Very local and usually coastal in the south of England and south Wales; scarce.
Similar species: Separated from *S. teutonus* (3) by its generally paler colour, the dark elytral mark being diffuse or absent; also differing in the pronotal side borders and basal punctuation.

3. *Stenolophus teutonus* (Schrank)

Length 5.5-6.5 mm. Head black, pronotum reddish brown, elytra red with a large black spot covering the apical two-thirds, except along the margins. Antennae black except for paler basal two segments and the extreme tip of the apical segment; legs yellow or pale brown. Pronotum iridescent, transverse, sides evenly rounded, side borders ending in front of the rounded hind angles; basal region unpunctured or with a few punctures in the foveae. Elytra moderately iridescent, intervals flat. On damp, open or disturbed ground near standing water; iv-vii. Very local in the south of England and south Wales; sometimes abundant.
Similar species: See *S. skrimshiranus* (2). Superficially resembles the larger species of *Badister* but these have two setiferous punctures inside each eye and truncated mandibles.

10. *ACUPALPUS* Latreille

Small beetles usually occurring in damp habitats. They are flatter than *Bradycellus* species, without a toothed mentum, and usually have the antennal segments darkened except at the base. Distinguished from *Bembidion* by not having reduced apical palpal segments. There are more than 100 species world-wide, with some 25 in Europe; one of the eight British species is apparently a recent immigrant. Only two species occur in Ireland. All are winged.

Key to species of *Acupalpus*

1. Pronotum uniformly black or dark brown 2

- Pronotum paler, at least at side margins 4

2. Basal third of elytra yellow; base of pronotum extensively punctured .... 7. *meridianus* (Linnaeus)

- Elytra entirely dark, except suture sometimes slightly paler; pronotal base with a few punctures in the foveae only 3

3. Length at least 3 mm; elytra without setiferous pore on the third interval .... 1. *brunnipes* (Sturm)

- Length less than 3 mm; third elytral interval with a minute setiferous pore in apical third (as Fig. 295) .... 4. *exiguus* Dejean

4. Length usually at least 4 mm; pronotum more transverse with hind angles completely rounded and not evident (Fig. 292); aedeagus with at least 12 internal teeth (Fig. 293) .... 3. *elegans* (Dejean)

- Length less than 4 mm; pronotum less transverse, hind angles rounded but distinct, sides in front of them straighter (Fig. 294); aedeagus with fewer or no internal teeth 5

5. Apical third of elytra with a minute pore on the third interval (Fig. 295) 6

- Apical third of elytra without any pore on the third interval .... 5. *flavicollis* (Sturm)

6. Length less than 3 mm; more or less uniformly mid brown with pronotum usually paler (Plate 120) .... 2. *dubius* Schilsky

- Length at least 3 mm; usually with distinct colour pattern, head black, pronotum bright red, or black with pale margins; elytra often bi-coloured 7

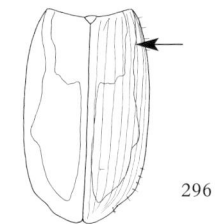

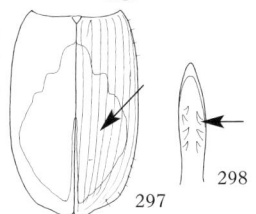

7. Base of elytra with a darkened region extending back along the sixth interval to join the main dark spot (Fig. 296); pronotum usually black with narrow pale borders; aedeagus without internal teeth .... 6. *maculatus* (Schaum)

- Base of elytra entirely pale or darkened only medially, usually with a dark spot in apical half (Fig. 297); pronotum entirely pale, or with smaller median black mark; aedeagus with fewer than 10 internal teeth (Fig. 298) .... 8. *parvulus* (Sturm)

1. *Acupalpus brunnipes* (Sturm)

Length 3-3.5 mm. Almost uniformly black or dark brown; only elytral suture slightly paler. Antennae dark with pale basal segment; legs pale brown. Pronotum very shiny, widest in front of middle, sides rounded throughout and hind angles hardly distinct; foveae with scattered punctures. Elytra without dorsal puncture on the third interval. In moss and litter near water, usually near the coast; iv-vii. Extremely local in southern England; very scarce.
Similar species: Distinguished from the other uniformly dark species, *A. exiguus* (4), by being larger, and lacking the dorsal elytral puncture on the third interval.

2. *Acupalpus dubius* Schilsky Plate 120

Length 2.5-2.8 mm. Pale to mid reddish brown, pronotum usually paler than head and much of elytra. Antennae dark brown with apex and first segment paler; legs yellow or pale brown, apices of tibiae sometimes darkened. Head narrower than pronotum. Pronotum widest well in front of middle, sides almost straight behind this, hind angles rounded but distinct; foveae almost impunctate. Elytra rather convex, dilated apically so that they are widest in apical third; third interval with dorsal puncture (Fig. 295). Wings present. In litter, moss and tussocks near fresh water; iv-viii. Widespread in southern Britain, local and often coastal in Wales and Ireland, with few records from northern England and Scotland; often very abundant.
Similar species: Paler than *A. exiguus* (4) the only other species of similar small size; distinguished from more or less uniformly pale examples of *A. flavicollis* (5) by the smaller size and presence of a puncture on the third elytral interval.

3. *Acupalpus elegans* (Dejean)

Length 4-4.5 mm. Head black, pronotum normally uniformly bright red-brown, elytra black with basal third, suture and margins red brown. Antennae dark with basal two segments pale; legs brown, femora slightly darker; male with both front and mid tarsi expanded. Pronotum transverse, sides strongly rounded and hind angles indistinct; foveae almost impunctate (Fig. 292). Elytra parallel-sided, rather truncate apically; dorsal puncture present on third interval. Aedeagus at least 1 mm long, with more than 12 internal teeth (Fig. 293). On salt marshes and under coastal cliffs in wet flushes; v-viii. Only known reliably from the Thames estuary, last recorded in 1952.
Similar species: Most resembles the pale pronotum form of *A. parvulus* (8) but larger, with more rounded pronotal sides. Larger and with more distinct elytral markings than *A. flavicollis* (5).

4. *Acupalpus exiguus* Dejean

Length 2.3-2.9 mm. Rather flat, uniformly black or very dark brown, or pronotum slightly paler. Antennae black or brown, base hardly paler; legs brown. Head almost as wide as pronotum. Pronotal sides distinctly contracted and straight in hind third, hind angles very rounded but distinct; base with a few punctures only in the shallow, indistinct foveae. Elytra finely microsculptured, third interval with dorsal puncture. In marshy sites with litter or tussocks, both inland and in salt marshes; iii-v. Very local and often coastal in southern England as far north as south Yorkshire and south Wales; can be abundant.
Similar species: See *A. brunnipes* (1) and *A. dubius* (2).

5. *Acupalpus flavicollis* (Sturm)

Length 2.8-3.5 mm. Head black, pronotum bright red-brown, elytra mid brown with suture and often base paler. Antennae dark brown with basal two segments pale; legs pale yellow-brown. Pronotum widest well in front of middle, sides behind this only slightly rounded, hind angles rounded but distinct; foveae without punctures. Elytra rather short, parallel-sided or slightly widened behind; third interval without any dorsal puncture, apices almost evenly rounded. In damp sandy and gravelly situations near water, also in lowland bogs; iv-vii. Very local in southern England; very scarce.
Similar species: Resembles the pale form of *A. parvulus* (8) but the elytra are less elongate, and lack a dorsal puncture on the third elytral interval. See also *A. dubius* (2) and *A. elegans* (3).

6. *Acupalpus maculatus* (Schaum)

Length 3.1-3.9 mm. Head black, pronotum usually dark with paler margins, elytra yellow-brown with a dark mark on apical third that extends forwards along the sixth interval to the elytral shoulders (Fig. 296). Antennae mid brown with basal segment paler; legs pale. Pronotum hardly transverse, widest at or just in front of middle, sides behind this rounded but hind angles evident; foveae hardly punctured. Elytra almost parallel-sided, third interval with dorsal puncture. Aedeagus less than 1 mm long, without internal teeth. In Britain only known since 1996 from Kent and Sussex in moss at the edges of pools and lakes.
Similar species: Most similar to the dark form of *A. parvulus* (8) but the pronotum is less transverse, and the dark markings on the pronotum and elytra are usually more extensive.

7. *Acupalpus meridianus* (Linnaeus)

Length 3-3.7 mm. Black or dark brown with basal third and suture of elytra pale yellow-brown; elytra very shiny, without microsculpture. Antennae dark brown with basal two segments paler; legs pale to mid brown, femora sometimes darker. Pronotum hardly transverse, sides widest in front of middle, almost straight and strongly contracted behind, hind angles rounded, base punctured throughout. Elytra long and rather narrow, striae punctured near base; third interval with a dorsal puncture. In gardens and fields on open clay or peat soils; iii-vii. Widespread in south-east England, local and often coastal further west and north as far as Yorkshire; abundant.
Similar species: Only dark forms of *A. parvulus* (8) have a similar colour pattern but in that species the margins of the pronotum are not dark. *A. meridianus* has shinier elytra than all other species of the genus, and also is the only one with a wholly punctured pronotal base.

8. *Acupalpus parvulus* (**Sturm**)

Length 3.1-3.9 mm. Very variable in colour but with two typical forms, 'pale' and 'dark'. Head black, pronotum either all red (pale) or with a variably-sized black median mark (dark); elytra red-brown with hardly any darker markings (pale), or with a distinct black patch (dark) covering at least apical third to half, and sometimes extending forwards medially to the scutellum but never laterally to the shoulders (Fig. 297). Antennae mid brown with basal one or two segments paler; legs pale or with apices of tibiae and bases of femora darker. Pronotum transverse, widest in front of middle, sides behind this almost straight, hind angles evident; foveae with scattered punctures (Fig. 294). Elytra almost parallel-sided, third interval with dorsal puncture. Aedeagus less than 1 mm long, with fewer than 10 internal teeth (Fig. 298). In damp habitats near vegetation; iv-viii. Widespread in England as far north as Yorkshire, local and coastal in Wales and southern Scotland; very local in south-west Ireland; can be abundant. The pale form is southern, the dark form more northerly in Britain.
Similar species: See *A. flavicollis* (5), *A. elegans* (3), *A. maculatus* (6) and *A. meridianus* (7).

11. ANTHRACUS Motschulsky

This genus was formerly included within *Acupalpus* but differs in having longer antennae, sinuate pronotal sides with sharp hind angles and conspicuous ventral pubescence. There are some 30 species world-wide but only two in Europe, one of which occurs in Britain and Ireland.

1. *Anthracus consputus* (**Duftschmid**) Plate 121

Length 3.8-5 mm. Rather flat; head black, pronotum and elytra mid-dark brown, with margins of pronotum, and margins, basal third and suture of elytra paler. Antennae very elongate, most segments three times as long as wide, black or dark brown with basal two segments pale; legs pale yellow-brown. Pronotum hardly transverse, widest well in front of middle, sides strongly contracted and sinuate in front of the sharp but obtuse hind angles; basal foveae deep, unpunctured. Elytra very long and parallel-sided or slightly widened apically; striae unpunctured; third interval with dorsal puncture in apical third. Wings present. Underside of prosternum and abdomen both pubescent. Under stones and in dense vegetation and litter near standing water; iv-viii. Widespread in southern and eastern England as far north as south Yorkshire, very local in the south-west of England, Wales and Ireland; scarce.
Similar species: Most resembles *Acupalpus meridianus* in shape and colouring but is larger with sharp pronotal hind angles, much longer antennae and pubescent underside. Superficially resembles *Blemus discus*, *Trechoblemus micros* or *Badister sodalis*.

Tribe CHLAENIINI

A large, mostly tropical tribe, also known as Callistini, with hundreds of medium to large-sized and often brightly-coloured predatory species. The British genera are characterised by the punctured and pubescent upper surface and single setiferous puncture inside each eye.

Key to genera of Chlaeniini

1. Length more than 8 mm; pronotum metallic or black (Plate 122); third antennal segment glabrous .... 1. *Chlaenius* Bonelli (p. 178)

- Length not more than 7 mm; pronotum yellow or orange, non-metallic (Plate 123); third antennal segment pubescent .... 2. *Callistus* Bonelli (p. 179)

1. *CHLAENIUS* Bonelli

This very large genus, with more than 30 European species, is sometimes split into smaller genera. Distinguished from *Callistus* by size, colouring and the lack of pubescence on the third antennal segment. Four species have been recorded in Britain, of which three also occur in Ireland. They are all winged.

Key to species of *Chlaenius*

1. Third antennal segment dark, usually black; legs at least partly black or dark brown 2

- Third antennal segment pale; legs entirely pale yellow-brown, or with femora and tarsi slightly darker 3

2. Body almost always brightly metallic coloured (Plate 122); first antennal segment brown, at least beneath .... 1. *nigricornis* (Fabricius)

- Body black with at most a slight metallic reflection on head and pronotum; antennae entirely black .... 3. *tristis* (Schaller)

3. Elytra with sides and apex pale yellow .... 4. *vestitus* (Paykull)

- Elytra uniformly green .... 2. *nitidulus* (Schrank)

1. *Chlaenius nigricornis* (Fabricius) Plate 122

Length 10-12 mm. Bright metallic; typically head coppery or green, pronotum reddish-coppery and elytra green; dark to almost black in some specimens. Antennae black with first segment paler, at least beneath; legs varying from all black (typical form), to red-brown with black or dark brown tarsi (var. *melanocornis* Dejean). Head very finely punctured. Pronotum as wide as elytra at shoulders, with strong rugose punctuation and long, narrow central furrow; sides rounded, not sinuate in front of hind angles. Elytra parallel-sided, densely punctured and finely pubescent. In damp grasslands and lowland marshes, also coastal litter; iv-viii. Widespread in England (except the extreme north) and Wales, local in Ireland; seldom abundant.
Similar species: Dark examples can be distinguished from *C. tristis* (3) by the pale underside of the first antennal segment and the wider pronotum relative to the elytra. Separated from *C. nitidulus* (2) by the black third antennal segment and the non-sinuate pronotal sides.

2. *Chlaenius nitidulus* (Schrank)

Length 10.5-12.5 mm. Bright metallic golden-green. Antennae dark brown with segments 1-3 pale; legs pale yellow-brown with femora and tarsi slightly darker. Head almost unpunctured. Pronotal sides contracted and slightly sinuate in front of hind angles, punctuation distinct, occasionally rugose. Elytra at shoulders wider than pronotum, widest behind middle, with dense punctuation and pubescence. In damp seepages under coastal cliffs; iv-vii. Only known from the coast of Dorset, Sussex and the Isle of Wight; last recorded in 1930.
Similar species: See *C. nigricornis* (1). Lacks the yellow elytral margins and apices seen in *C. vestitus* (4).

3. *Chlaenius tristis* (Schaller)

Length 11-13 mm. Black, sometimes with a coppery metallic reflection. Appendages entirely black. Pronotum narrow relative to elytra, sides not sinuate. Elytra rather long and convex, pubescence reddish-golden. In densely vegetated bogs and flushes; iv-viii. Formerly known from the East Anglian Fens but since 1900 only recorded from the Llyn peninsula, North Wales, and Kerry and Westmeath in Ireland; extremely scarce.
Similar species: See *C. nigricornis* (1).

4. *Chlaenius vestitus* (Paykull)

Length 9-11 mm. Bright green, sometimes with a golden or bluish tint; sides and apical section of elytra pale yellow. Appendages pale. Head almost unpunctured. Pronotum small and narrow, hardly wider than head; sides sinuate in front of sharp hind angles; punctuation dense only near margins. Elytra widened behind, densely punctured and with pale pubescence. In mud and clay cracks near water; iv-viii. Widespread in south and east England as far north as Yorkshire, also south Wales; very local in Ireland and not recorded since 1938; scarce.
Similar species: Distinguished from all other species by the pale elytral margins.

2. *CALLISTUS* Bonelli

A single, very distinctive, European species.

1. *Callistus lunatus* (Fabricius) Plate 123

Length 6-7 mm. Head blue-black, pronotum orange; elytra pale yellow, each with black shoulder, a median spot and apical transverse band. Entire upper surface finely punctured and pubescent. Antennae black with basal two segments pale and non-pubescent; legs finely pubescent, pale with most of femora, tibial apices and tarsi darkened. Pronotum cordate, hind angles rectangular. Elytra rounded and convex. Wings present. On open chalk downland; v-vi. Extremely local in south-east England, and not recorded since 1953.

Tribe OODINI

A rather diverse tribe of more than 200 hygrophilous species world-wide but with only four European species in the single genus *Oodes*, of which one occurs in Britain but not in Ireland. Characterised by the outermost elytral stria forming an apical keel around the hind margins of the elytra (Fig. 41); the head has a single puncture inside each eye and the third antennal segment is glabrous.

1. *Oodes helopioides* (Fabricius) Plate 124

Length 7.5-10 mm. Body outline very smooth; uniformly black, dorsal surface finely microsculptured and not very shining; apex of mandibles and first antennal segment sometimes a little paler. Pronotum very transverse, as wide as elytra, not punctured, hind margin concave, hind angles rectangular but rounded. Elytra wide and parallel-sided, striae finely punctured; third interval with two dorsal punctures in apical half; outermost (eighth) stria continuing around the apex of each elytron and forming a keel running above the true hind margin. Wings present. In litter and vegetation at the edges of lakes, slow rivers and fens, sometimes submerged in the water; iv-viii. Widespread but local in southern England and south Wales, occasionally further north to Yorkshire and Cumbria; can be abundant.

Similar species: There is a superficial resemblance to a large, dark *Amara* but *Oodes* has only a single seta inside each eye, and no setae on the palpi.

Tribe LICININI

A small tribe of specialised snail-eating carabids, characterised by a modified labrum and asymmetrical, truncated mandibles (Fig. 30). The two British genera are sometimes put into separate tribes, Licinini and Badistrini.

Key to genera of Licinini

1. Length at least 9 mm; black species with punctures on elytral intervals .... 1. *Licinus* Latreille (p. 180)

- Length usually less than 9 mm; dark or brightly coloured species with elytral intervals unpunctured .... 2. *Badister* Clairville (p. 181)

1. *LICINUS* Latreille

There are about 20 Palaearctic species in this genus, with 14 in Europe but only two in Britain. Neither of these occur in Ireland. Easily recognised by the large size, black colour and extensive punctuation. Males have the apical palpal segments enlarged and hatchet-shaped, as well as very swollen two basal front tarsal segments.

Key to species of *Licinus*

1. Length usually less than 11 mm; elytral surface flat, intervals each with two or three irregular rows of fine punctures (Fig. 299) .... 1. *depressus* (Paykull)

- Length usually more than 11 mm; elytral surface uneven, intervals each with a single row of larger punctures (Fig. 300) .... 2. *punctatulus* (Fabricius)

1. *Licinus depressus* (Paykull) Plate 125

Length 9-11.5 mm. Body and appendages entirely black; head and pronotum shining, elytra duller, especially in females. Head often very large relative to pronotum, punctured especially towards the rear. Pronotum moderately transverse, widest in front of middle, sides evenly rounded, punctured throughout. Elytra usually no wider than pronotum, shoulders angulate, surface smooth with finely punctured striae and densely but finely punctured intervals (Fig. 299); apices obliquely truncate and moderately sinuate. Wings absent. In both open and shaded habitats on dry, sandy or calcareous soils, also in gravel pits; vii-xi. Local in south and east England as far north as Co. Durham, also in south Wales; scarce.

2. *Licinus punctatulus* (Fabricius)

Length 11.5-16 mm. Body and appendages entirely black; head and pronotum shining, elytra slightly duller. Head much narrower than pronotum, finely punctured. Pronotum very transverse, punctured densely towards the margins but more finely on the disc. Elytra wide, with shoulders rounded, intervals 3, 5 and 7 usually somewhat raised; all intervals with an irregular row of deep punctures in transverse depressions (Fig. 300); apices oblique and strongly sinuate. Wings present. On dry, chalky or sandy soils; vii-x. Extremely local in southern England, usually on the coast; very scarce.

2. *BADISTER* Clairville

This world-wide genus has about 70 species, sometimes put in their own tribe Badistrini. They are smaller than *Licinus*, with unpunctured elytral intervals and unmodified palpi. Seven of the 10 European species occur in both Britain and Ireland. Their identification can require dissection of the aedeagus, especially in the smaller species (subgenus *Baudia*).

Key to species of *Badister*

1. Pronotum bright red-brown, elytra reddish with black markings (Plate 126) 2

- Pronotum dark brown or black with only margins paler; elytra mainly or entirely dark brown or black 4

2. Scutellum pale, usually reddish; head almost as wide as pronotum, which is strongly contracted behind (Fig. 301) ....
.... 3. *unipustulatus* Bonelli

- Scutellum darker than pronotum and elytral base, often black; head narrower than pronotum, which is less contracted behind (Plate 126) 3

3. First antennal segment pale, with at most a darker shadow at extreme apex; aedeagal tip protruding both above and below (side view Fig. 302) .... 1. *bullatus* (Schrank)

- First antennal segment darkened, at least apically; aedeagal tip protruding below only (side view Fig. 303) ..... 2. *meridionalis* Puel

4. Elytra with distinct paler spot behind the shoulder (Plate 127); right mandible with a dorsal notch (Fig. 304) ....
.... 4. *sodalis* (Duftschmid)

- Elytra unicolorous brown; left mandible notched (Fig. 305) 5

5. Length at least 5 mm; pronotal hind angles more rounded (Plate 128); mandibles more curved apically (Fig. 305); aedeagal apex with a ventral hook well-removed from, and pointing away from, apex (side view Fig. 306) .... 6. *dilatatus* Chaudoir

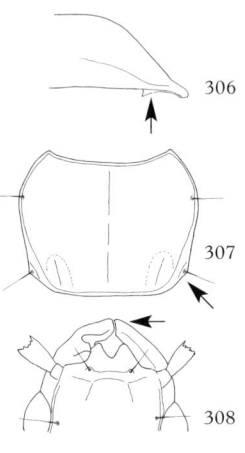

- Length usually less than 5 mm; pronotal hind angles obtuse but more distinct (Fig. 307); mandibles less curved apically (Fig. 308); aedeagal apex not as above 6

6. Striae very fine, intervals completely flat; aedeagal apex with hook just before apex (side view Fig. 309) .... 5. *collaris* Motschulsky

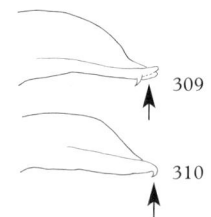

- Striae deeper, intervals very slightly convex; aedeagal apex with apical hook (side view Fig. 310) .... 7. *peltatus* (Panzer)

Subgenus *Badister sensu stricto*

1. *Badister bullatus* (Schrank) Plate 126

Length 4.8-6.3 mm. Head black, pronotum bright red-brown, scutellum black, elytra red with median and apical black spots that are widely joined along the side margins, leaving the sutural striae pale. Antennae dark with first segment red, and apical two or three segments gradually paler; legs pale red-brown with tarsi darkened. Head narrower and shorter than pronotum. Pronotal sides rounded, only slightly contracted to the rounded hind angles. Elytral sides rounded, dorsal surface hardly iridescent. Aedeagus protruding both ventrally and (slightly) dorsally at apex (Fig. 302). Wings present. *B. bullatus* appears to include two distinct taxa, whose specific status is not yet clear. One 'form' is on average larger, with more extensive dark elytral markings; it is found in most habitats, especially open, dry often sandy situations such as lowland heaths, grasslands and dunes. Widespread throughout Britain and Ireland but coastal towards the north; iv-viii; abundant. The second 'form' is on average smaller, with elytral markings not extending so far forwards; it has a more restricted distribution in south and east England, occurring mainly along the banks of lowland rivers.
Similar species: Distinguished from *B. meridionalis* (2) by the completely pale first antennal segment; it is also generally shorter than that species, with more rounded elytra, a smaller head and less iridescent elytra. The dark scutellum (and mesepisterna) separates both these species from *B. unipustulatus* (3) which also has the pronotum more strongly contracted behind.

2. *Badister meridionalis* Puel

Length 5.8-7.2 mm. Dorsal surface similar to *B. bullatus*. Antennae varying from entirely black, to dark with basal half to two-thirds of first segment red, and apical two or three segments gradually paler; legs pale with tarsi often darkened. Head almost as wide as, and as long as, the pronotum. Elytra iridescent, sides almost parallel. Aedeagus with apex deflected ventrally only (Fig. 303). Wings present. Under stones and in grass tussocks near water. Recorded only from Oxfordshire and Gloucestershire in England and from Galway in Ireland; very scarce.
Similar species: See *B. bullatus* (1).

3. *Badister unipustulatus* Bonelli

Length 6.8-9 mm. Head black, pronotum and scutellum bright red-brown; elytra very strongly iridescent, red with long median and short apical black spots that are usually joined along the side margins but always leaving the sutural striae pale. Antennae with apex of first segment and segments 2-6 darkened, usually black, apical segments gradually paler; legs pale yellow-brown with tarsi darkened. Head very large, as wide as pronotum. Pronotum widest just behind front angles, sides rounded and strongly contracted to indistinct hind angles (Fig. 301). Elytra long, usually widest behind middle. Wings present. In marshes and damp litter in well-vegetated sites near water, iv-viii. Very local in south and east England as far north as Yorkshire, with a single Welsh record; only Clare and Kerry in Ireland; scarce
Similar species: See *B. bullatus* (1).

Subgenus *Trimorphus* Stephens

4. *Badister sodalis* (Duftschmid) Plate 127

Length 3.8-4.6 mm. Head black, pronotum dark brown or black with margins paler, elytra mid to dark brown with a paler spot behind the shoulders, extending up to one third of the elytral length and in as far as stria 2. Antennae black with paler apical segments; legs pale yellow-brown. Right mandible with a dorsal notch (Fig. 304). Pronotum contracted and very slightly sinuate in front of the rounded hind angles. Elytral sides rounded, widest behind middle, intervals convex basally. Wings present or absent. In litter in damp woodlands, usually on heavy soils; iv-vii. Widespread in England except the north, local in Wales, south-west Scotland and Ireland; scarce.
Similar species: Similar in size to small examples of *B. collaris* (5) and *B. peltatus* (7) but distinguishable by the pale spot at the base of each elytron and the right (rather than left) notched mandible.

Subgenus *Baudia* Ragusa

5. *Badister collaris* Motschulsky

Length 4.2-5 mm. Head black, pronotum and elytra dark brown to black with paler margins. Antennae black, sometimes paler apically and at base; legs brown with tarsi often darker. Left mandible with a dorsal notch, hardly curved apically (Fig. 308). Pronotum strongly contracted to the obtuse but distinct hind angles. Elytral striae very fine, intervals flat or nearly so. Aedeagus with sub-apical hook just behind the apex (Fig. 309). Wings present. In litter at the well-vegetated edges of ponds and flooded gravel pits; iv-vii. Extremely local in the extreme south of England and at a single Irish site in Clare; very scarce.
Similar species: On average smaller than *B. dilatatus* (6) with more distinct pronotal hind angles. Elytral intervals usually flatter than *B. peltatus* (7) and pronotum slightly less contracted at base. Separable with certainty only by the aedeagus. See also *B. sodalis* (4).

6. *Badister dilatatus* Chaudoir Plate 128

Length 4.5-5.4 mm. Black or dark brown, margins hardly paler. Antennae black; legs mid to dark brown. Left mandible with a dorsal notch, strongly curved apically (Fig. 305). Pronotum moderately contracted, hind angles almost completely rounded and indistinct. Elytra wide, widest behind middle, striae very fine, intervals flat or nearly so. Aedeagus

with sub-apical hook well before the apex (Fig. 306). Wings present. In litter and vegetation near standing water; iv-vii. Local in south and east England, and the coasts of south-west England, Wales and southern Ireland; scarce.
Similar species: Larger on average than the other species of subgenus *Baudia*, with more rounded pronotal hind angles. Separable with certainty only by the aedeagus.

7. *Badister peltatus* (Panzer)

Length 4.3-5 mm. Head black, pronotum and elytra dark brown to black with paler margins. Antennae black, sometimes paler apically and at base; legs brown with tarsi often darker. Left mandible with a dorsal notch, hardly curved apically. Pronotum strongly contracted to the obtuse but distinct hind angles (Fig. 307). Elytral intervals slightly convex, striae deepened basally. Aedeagus with apical hook (Fig. 310). Wings present. In litter and vegetation near standing water, usually coastal; iv-vii. Very local in southern England and Wales, with a few inland records; very scarce. Also in south-west Ireland; scarce.
Similar species: See *B. sodalis* (4), *B. collaris* (5) and *B. dilatatus* (6).

Tribe PANAGAEINI

Both European species of this tribe are in the genus *Panagaeus*, which has about 13 species distributed throughout the Holarctic. They have a densely punctured and pubescent upper surface, protruding eyes and characteristic black and red elytral patterns. There are two setiferous punctures inside the eyes, although these can be difficult to distinguish among the dense general pubescence. The third antennal segment is pubescent. The palpi have expanded apical segments that are inserted obliquely on the end of the preceding segments. Both British species are winged; only one is found in Ireland.

Key to species of *Panagaeus*

1. Pronotum slightly wider than long (Fig. 311) with small and large punctures; hind red spot on elytra not reaching the side margin .... 1. *bipustulatus* (Fabricius)

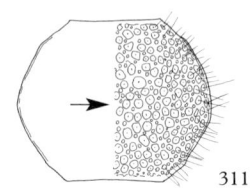

- Pronotum much wider than long with large punctures only; both red spots on elytra usually reaching the side margin (Plate 129) .... 2. *cruxmajor* (Linnaeus)

1. *Panagaeus bipustulatus* (Fabricius)

Length 6.5-7.5 mm. Head and pronotum black, elytra black with two large red spots on each, leaving only the inner two intervals and the outermost interval outside the rear spot black. Appendages black, or paler apically. Head finely punctured at hind margin only, eyes almost hemispherical. Pronotum slightly wider than long (Fig. 311), densely punctured and pubescent, with fine punctures between the larger pores. Elytra pubescent, striae coarsely punctured, intervals convex. On open, well-drained grasslands and dunes, also chalk and gravel pits; iv-vii. Local in England except the north, and south Wales; scarce.
Similar species: Distinguished from *P. cruxmajor* (2) by the pronotum, and smaller hind red spot on the elytra.

2. *Panagaeus cruxmajor* (Linnaeus) Plate 129

Length 7.4-8.8 mm. Head and pronotum black, elytra black with two large transverse red spots on each, leaving only the innermost interval black. Appendages black. Head finely punctured at hind margin only, eyes protruding and hemispherical. Pronotum much wider than long, densely punctured with large pores only, and pubescent. Elytra pubescent, striae moderately punctured, intervals flat. In litter, well-vegetated fens and dune slacks, always near water; v-vi. Widely distributed but extremely local in England as far north as Lincolnshire, south Wales and south-west Ireland; usually very scarce.
Similar species: See *P. bipustulatus* (1).

Tribe PERIGONINI

This tropical and sub-tropical tribe has about 200 species, with about 80 in the large genus *Perigona* Laporte. They are small, delicate and sometimes pubescent carabids characterised by the deepened outer elytral stria, which continues around to the elytral apex. There is a single, cosmopolitan British species, probably introduced early in the twentieth century.

1. *Perigona nigriceps* Dejean Plate 130

Length 2-2.5 mm. Head black except in front, remainder of body pale yellow or light brown, with apex and sides of elytra darkened; appendages pale yellow. Mandibles strongly protruding forwards; eyes large and prominent; frontal furrows absent. Pronotum widest near front, sides curved and contracted to obtuse and rounded hind angles. Elytra with striae reduced to fine puncture rows, except deepened eighth stria; finely pubescent apically and at sides. Wings present. In compost and grass heaps; iv-x. Widespread but sporadic and local in England as far north as Yorkshire, and south Wales; usually abundant.
Similar species: Superficially resembles *Trechus quadristriatus* and *T. obtusus* in size, shape and colouring but distinct by the lack of frontal furrows and deepened eighth elytral stria.

Tribe MASOREINI

This mainly tropical tribe has a single European species in the genus *Masoreus* Dejean. It is characterised among those carabids with truncate elytra by having spiny tibiae (Fig. 42) and a very wide pronotum.

1. *Masoreus wetterhallii* (Gyllenhal) Plate 131

Length 4.5-6 mm. Head and apical two-thirds of elytra dark shining reddish-brown, pronotum and elytral base mid brown. Appendages light brown; antennae with three basal segments glabrous; tibiae with stout spines externally. Pronotum very wide, widest in front of middle, sides curved in to obtuse, rounded hind angles; base slightly sinuate and protruding medially. Elytra wide and short, rounded at sides; striae almost impunctate, third interval with two dorsal punctures; apices rounded externally and obliquely truncate, exposing only the apex of the last abdominal tergite. Wings usually absent. In sand and gravel soils, usually near the coast; often under low vegetation; v-viii. Very local around the south and east coasts of England from Cornwall to Lincolnshire; also inland in Norfolk and Suffolk; very scarce.
Similar species: Superficially resembles some *Trechus* but the head has no deep furrows and the truncate elytra do not have a recurrent sutural stria.

Tribe LEBIINI

This is a large and very diverse tribe, with truncate elytra and narrow, bordered pronotum. The Demetriini and Dromiini, sometimes considered as separate tribes, are included. There are over 100 European species, with many more in tropical regions. Ten genera are recognised in the British and Irish fauna; the former genus *Dromius* is now divided into four genera. Two further species, *Somotrichus unifasciatus* (Dejean) and *Plochionus pallens* (Fabricius), are occasionally introduced into Britain on stored products such as Brazil nuts, but have not occurred outside the ports of importation and have never become established.

Key to genera of Lebiini

1. Elytra punctured throughout and usually pubescent; length usually more than 8 mm .... 3. *Cymindis* Latreille (p. 190)

- Elytra if punctured, not pubescent except at extreme margins; length at most 8 mm 2

2. Pronotum very transverse, centre of hind margin protruding abruptly backwards (Plates 132, 133) .... 1. *Lebia* Latreille (p. 187)

- Pronotum not or less transverse, hind margin more or less straight ... 3

3. Tarsi with penultimate segment bilobed (Fig. 312) .... 2. *Demetrias* Bonelli (p. 188)

- Tarsi simple, not bilobed 4

4. Antennae entirely black, or with first segment sometimes paler 5

- Antennae pale, usually reddish, rarely brown 7

5. Elytral intervals with a row of fine punctures between the striae (Fig. 313); pronotum strongly contracted behind, hind angles reduced to small teeth (Plate 143) .... 9. *Lionychus* Wissmann (p. 197)

- Elytral intervals unpunctured, or with a few deeper pores on third interval only; pronotum less contracted, hind angles obtuse but sharp and distinct (Fig. 314) 6

6. Elytral apices transversely truncate (Fig. 315); third antennal segment pubescent .... 8. *Microlestes* Schmidt-Göbel (p. 196)

- Elytral apices obliquely truncate (Fig. 316); third antennal segment glabrous .... 10. *Syntomus* Hope (p. 197)

7. Body slender, with quadrate or elongate pronotum, elytra together at least twice as long as wide (Plate 138) .... 4. *Paradromius* Fowler (p. 191)

- Body broader, pronotum transverse, elytra relatively less elongate (Plates 139-141) 8

8. Length at least 5 mm .... 5. *Dromius* Bonelli (p. 192)

- Length less than 5 mm 9

9. Pronotum quadrate, hind angles protruding; raised basal border of elytra almost reaching the scutellum (Fig. 317) .... 6. *Calodromius* Reitter (p. 194)

- Pronotum transverse, hind angles not protruding; raised basal border of elytra incomplete, not nearly reaching the scutellum (Fig. 318) .... 7. *Philorhizus* Hope (p. 194)

1. *LEBIA* Latreille

Separable from the other taxa with truncate elytra by their bright colours, the wide pronotum which protrudes back medially and widened fourth tarsal segment. *Lebia* is an enormous, mainly tropical genus with over 450 species. Their larvae are ectoparasitoids of chrysomelid larvae and pupae, while the adults can be found on vegetation and in tussocks. Only three species occur in Britain, two in Ireland; two further species, *L. marginata* (Fourcroy) and *L. scapularis* (Fourcroy), have not been found since the nineteenth century and are not included in the key. All are fully winged.

Key to species of *Lebia*

1. Elytra uniformly metallic blue or green 2

- Elytra black with large pale yellow markings (Plate 133) .... 3. *cruxminor* (Linnaeus)

2. Femora entirely pale yellow-brown .... 1. *chlorocephala* (Hoffmann, J.)

- Femora with apical third darkened, usually black .... 2. *cyanocephala* (Linnaeus)

Subgenus *Lamprias* Bonelli

1. *Lebia chlorocephala* (Hoffmann, J.) Plate 132

Length 5.8-8 mm. Head and elytra bright metallic golden- or bluish-green, pronotum reddish yellow. Antennae black with segments 1-2 (and usually base of 3) pale; legs pale with tarsi darkened. Pronotum widest well in front of middle, with scattered punctures. Elytra wide and rounded at sides, striae punctured more strongly than the finely punctured intervals; margins with well-separated very deep pores. In open grasslands, especially where there are tussocks, usually associated with *Chrysolina* species; iii-v. Widespread in southern England, more local in the north, coastal in Wales and eastern Scotland, very occasional in Ireland; seldom abundant.
Similar species: *L. cyanocephala* (2) has a bluer head and elytra, with darker appendages and stronger punctures on the elytral intervals.

2. *Lebia cyanocephala* (Linnaeus)

Length 5-7.5 mm. Head and elytra bright metallic blue, pronotum orange. Antennae black with first segment pale; legs black with base of femora and middle of tibiae paler. Pronotum widest slightly in front of middle, evenly punctured. Elytra with sides straight, diverging to near apices; intervals punctured more strongly than the fine striae; outer intervals pubescent, partly obscuring the marginal pores. In dry sandy grasslands and heaths, probably in association with *Chrysolina* species; iii-v. Extremely local in southern England, the only recent records are from Surrey; very scarce.
Similar species: See *L. chlorocephala* (1).

Subgenus *Lebia sensu stricto*

3. *Lebia cruxminor* (Linnaeus) Plate 133

Length 5-7 mm. Head black, pronotum orange-red, elytra yellow with black markings forming an approximate cross. Antennae black with basal three segments pale; legs pale with femoral apices and tarsi blackened. Pronotum almost unpunctured. Elytra extremely wide and convex, sides rounded, widest near apex; intervals extremely finely punctured. In grasslands, probably associated with *Galeruca tanaceti* (Linnaeus); iv-vi. Very local, with scattered recent records from Cornwall and Sussex in England, and Fermanagh, Clare and Kerry in Ireland; very scarce.

2. *DEMETRIAS* Bonelli

This small genus has about ten Palaearctic species of which three occur in Britain, one in Ireland. They are distinguished from *Dromius* and related genera by their bilobed fourth tarsal segment (Fig. 312).

Key to species of *Demetrias*

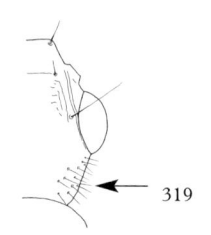

1. Head punctured and pubescent behind eyes (Fig. 319) ....
   .... 2. *atricapillus* (Linnaeus)

- Head smooth behind eyes 2

319

2. Elytra with a single median dark spot near apex (Fig. 320); wings absent .... 3. *monostigma* Samouelle

- Elytra with three dark spots, these sometimes joined (Plate 134); wings present .... 1. *imperialis* (Germar)

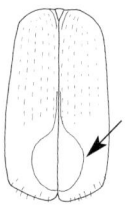

320

Subgenus *Risophilus* Leach

1. *Demetrias imperialis* (Germar) Plate 134

Length 4.5-5.8 mm. Body generally flat. Head black, pronotum red-brown, elytra pale straw-coloured with a median diamond-shaped dark mark which is widest in apical third, and a lateral dark spot on each elytron just behind this. Appendages entirely pale, or antennae slightly darkened from third segment to apex; claws without teeth. Head without punctures or setae behind the eyes. Pronotum very narrow, without foveae inside the sharp hind angles. Third elytral interval with four punctures. Wings present, visible through the elytra. In reed beds and associated litter; iv-vi. Local in south and east England; scarce.
Similar species: Differs from the other species of *Demetrias* by having simple claws and no fovea inside the pronotal hind angles; the characteristic elytral pattern is usually sufficient to recognise this species.

Subgenus *Demetrias sensu stricto*

2. *Demetrias atricapillus* (Linnaeus) Plate 135

Length 4.5-5.5 mm. Head black, pronotum orange-red, elytra straw-coloured with a faint darker sutural mark that widens apically. Appendages pale; claws with three small teeth. Head with scattered punctures and pubescence behind the eyes, and laterally on the temples (Fig. 319). Pronotum with a small deep fovea just inside each hind angle. At least elytral intervals 3 and 5 with rows of fine pores. Wings present, visible through the elytra. On dunes, in tussocky grasslands and agricultural fields; iii-vi. Widespread in south and east England, local and usually coastal in the west (including Wales), Isle of Man and Ireland; very abundant.
Similar species: Distinguished from the other species of *Demetrias* by the almost unmarked elytra and the pubescent hind region of the head.

3. *Demetrias monostigma* Samouelle

Length 4.5-5.5 mm. Head black, pronotum and elytra pale red-brown, the latter with a thin dark sutural line expanding to form a single sub-apical spot (Fig. 320). Appendages pale; claws each with a single tooth. Pronotum with a small deep fovea just inside each hind angle. Elytra narrow, especially at the rounded shoulders, widening slightly towards apex; third interval with three or four fine punctures. Wings absent. In litter and on vegetation in wet dunes and fens; iii-vi. Local and often coastal in England from Yorkshire to south Wales; scarce.
Similar species: More slender than the other species of *Demetrias* and distinguished from both by the lack of wings. The body shape resembles *Paradromius linearis* but that genus does not have bilobed tarsi.

3. *CYMINDIS* Latreille

This large Holarctic genus has about 200 species, of which more than 40 occur in Europe. The British species are characterised by their punctured and often pubescent elytra and pectinate claws. The shape of the head and pronotum differ from *Polistichus* (tribe Zuphiini) which is also punctured and pubescent. One of the three British species is a recent discovery and only one occurs in Ireland.

Key to species of *Cymindis*

1. Elytral intervals each with an irregular row of punctures but not pubescent (Plate 136) .... 1. *axillaris* (Fabricius)

- Elytral intervals densely punctured and finely pubescent 2

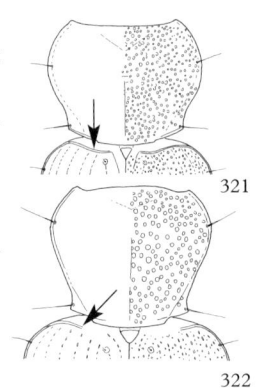

2. Raised basal border of elytra extending to meet the base of the first stria (Fig. 321); pronotal hind angles obtuse .... 2. *macularis* Mannerheim

- Raised basal border of elytra abbreviated level with the third stria (Fig. 322); pronotal hind angles right-angled .... 3. *vaporariorum* (Linnaeus)

Subgenus *Cymindis sensu stricto*

1. *Cymindis axillaris* (Fabricius) Plate 136

Length 8-10.5 mm. Head almost black, pronotum red-brown, elytra black or very dark brown with yellow outer margins and a shoulder mark covering the base of intervals 5-7; rarely this extends as a pale curved streak almost to elytral apex. Appendages pale to mid reddish brown; tarsal claws finely toothed. Head and pronotum finely punctured. Pronotum with hind angles hardly protruding. Elytral basal border complete; intervals slightly convex, very shining, each with a single irregular row of punctures; apices obliquely truncate, slightly sinuate. Wings usually present. On sandy heaths, shingle banks and dry grasslands; iv-vii. Very local in south east England; often coastal but recently mainly inland in Norfolk and Suffolk; scarce.
Similar species: *C. macularis* (2) also has a distinct shoulder mark and complete elytral basal border but its head and pronotum are more densely punctured, and the elytra are pubescent.

2. *Cymindis macularis* Mannerheim

Length 7-8.5 mm. Almost uniformly dark reddish brown, head darkest, elytra with a pale shoulder mark from interval 6 outwards. Appendages yellow-brown. Head and pronotum densely and strongly punctured. Pronotal hind angles obtuse, not protruding. Elytral base completely bordered (Fig. 321); intervals flat, densely punctured and pubescent; apices obliquely rounded, not sinuate. Wings probably present or absent. In dry lowland heaths;

vii-x. Only known in Britain since 1966 from Suffolk; very scarce.
Similar species: See *C. axillaris* (1). Distinguished from *C. vaporariorum* (3) by having a completely bordered elytral base, and a more distinct pale shoulder mark on the elytra.

Subgenus *Tarulus* Bedel

3. *Cymindis vaporariorum* (Linnaeus) Plate 137

Length 8-10 mm. Black or very dark red-brown, basal third of elytra paler, dark reddish. Appendages mid-brown. Head and pronotum densely and strongly punctured. Pronotum hardly transverse, hind angles protruding, rectangular. Raised basal border of elytra ending in front of the third elytral interval, not nearly reaching scutellum (Fig. 322). Elytral intervals each with two or three irregular rows of punctures and pubescence; apices obliquely truncate, not sinuate. Wings present or absent. Usually in upland bogs and wet moorland with *Calluna*; also lowland sandy heaths if kept damp by rainfall; v-viii. Widespread but local in northern England and Scotland; occasionally in north Wales and Ireland; scarce.
Similar species: See *C. macularis* (2).

4. *PARADROMIUS* Fowler

A small Palaearctic genus of about 15 species, formerly part of *Dromius*. Distinguished from *Demetrias* by the non-bilobed tarsi, and from *Dromius* and *Philorhizus* by their more elongate body shape. They occur on, or in the stems of, grasses and reeds.

Key to species of *Paradromius*

1. Head (excluding mandibles) hardly longer than wide, frons with several longitudinal furrows inside the eyes (Plate 138) ....
   .... 1. *linearis* (Olivier)

- Head much longer than wide, frons with at most a single furrow inside each eye (Fig. 323) .... 2. *longiceps* (Dejean)

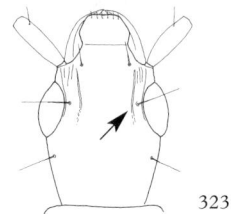

1. *Paradromius linearis* (Olivier) Plate 138

Length 4.5-5.5 mm. Head and pronotum red-brown, elytra yellow-brown, slightly darker apically and along suture. Appendages pale; antennae with three basal segments glabrous. Head (excluding mandibles) as long as wide, frons strongly wrinkled inside the eyes; temples as long as longitudinal diameter of eyes. Pronotum quadrate, side borders hardly raised. Elytra together 2.5 times as long as wide, widest near apex, with distinct punctured striae; base bordered from stria 3 outwards. Wings usually absent, rarely present. In dry grasslands, arable fields and dunes; iii-viii. Widespread in Britain and Ireland, mainly coastal in the north; very abundant.
Similar species: Less elongate than *P. longiceps* (2) which usually has a dark median mark towards the apex of the elytra.

2. *Paradromius longiceps* (Dejean)

Length 5-6.5 mm. Head dark brown, pronotum and elytra red-brown, elytra with a dark sutural mark in apical third, this sometimes expanded almost to reach sides of elytra. Appendages pale, all very elongate. Antennae with only two basal segments glabrous. Head longer than wide, temples much longer than longitudinal diameter of the eyes; frons with only a single furrow inside each eye (Fig. 323). Pronotum elongate, side borders raised. Elytra three times as long as together wide, widened to apex, with shoulders completely rounded, bordered at base from stria 3 outwards; striae very fine but punctured. Wings present. In reed beds and coastal litter; iv-vi. Very local in eastern England, mainly Yorkshire and East Anglia; very scarce.
Similar species: See *P. linearis* (1).

5. *DROMIUS* Bonelli

In its present, strict sense, the genus comprises about 10 European species found on trees where they are predatory on and under the bark. They are flat, with very faint striae, untoothed claws and an incomplete basal border on the elytra. Larger than species of *Philorhizus* and *Calodromius*. There are four British species, two of which also occur in Ireland; all are winged.

Key to species of *Dromius*

1. Elytra black each with two well-defined pale marks (Plate 139) .... 4. *quadrimaculatus* (Linnaeus)

- Elytra mid to dark brown, sometimes with diffuse paler area in front third 2

2. Third elytral interval with about five shallow and fine setiferous punctures (Fig. 324) .... 1. *agilis* (Fabricius)

- Third elytral interval with an apical puncture only 3

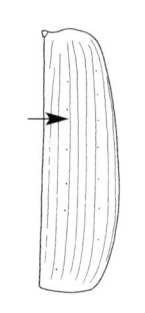

3. Pronotum less transverse, with narrower side borders; frons wrinkled just inside the eyes only (Fig. 325) ...... 2. *angustus* Brullé

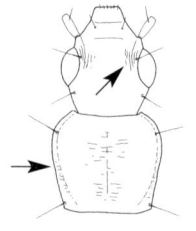

- Pronotum more transverse, with wide raised side borders; frons extensively wrinkled (Fig. 326) .... 3. *meridionalis* Dejean

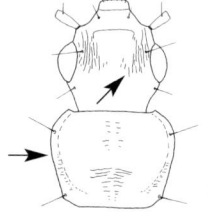

1. *Dromius agilis* (Fabricius)

Length 5.5-6.8 mm. Mid to dark reddish brown, sometimes with an indistinct pale mark near the base of each elytron. Appendages pale. Frons with longitudinal and oblique wrinkles inside each eye. Pronotum only slightly wider than long, with wide raised side borders that are hardly narrowed in front. Elytra together less than twice as long as wide, striae very fine; third and seventh intervals each with at least five small setiferous punctures (Fig. 324). In both coniferous and deciduous woodlands; v-ix. Widespread in Britain; seldom abundant.

Similar species: Separated from *D. angustus* (2) and *D. meridionalis* (3) by the fine punctures on the third elytral interval. *D. agilis* has a more distinctly transverse pronotum than either of these species, and relatively shorter elytra and a rougher frons than *D. angustus*.

2. *Dromius angustus* Brullé

Length 5.5-6.8 mm. Mid to dark reddish brown, usually with an indistinct pale mark near the base of each elytron. Appendages pale. Frons usually wrinkled inside only the front part of each eye and smooth medially (Fig. 325). Pronotum small relative to elytra, only slightly wider than long, with raised side borders rather narrowed in front. Elytra together more than twice as long as wide, striae clearly punctured; seventh interval with small setiferous punctures. In coniferous woodlands; v-x. Very local, mainly in Eastern Britain; scarce.

Similar species: See *D. agilis* (1). More elongate than both that species and *D. meridionalis* (3) and with a less wrinkled frons than *D. meridionalis*, which also has a more transverse pronotum.

3. *Dromius meridionalis* Dejean

Length 5.5-6.8 mm. Mid to dark reddish brown, sometimes with an indistinct pale mark near the base of each elytron. Appendages pale. Frons usually wrinkled inside much of each eye, these wrinkles extending over most of the frons (Fig. 326). Pronotum distinctly transverse, with raised side borders rather narrowed in front. Elytra together less than twice as long as wide, striae very finely punctured; seventh interval with small setiferous punctures. Mainly in deciduous woodlands; v-x. Widespread in England and Wales, local in Scotland and Ireland; can be abundant.

Similar species: See *D. agilis* (1) and *D. angustus* (2).

4. *Dromius quadrimaculatus* (Linnaeus) Plate 139

Length 5-6 mm. Head dark brown, pronotum bright orange-brown (sometimes darker centrally), elytra black with two pale straw-coloured spots on each; the hind spot includes the entire hind margin of each elytron. Appendages very pale yellow. Frons covered in fine longitudinal wrinkles. Pronotum transverse, with strongly raised side borders. Elytral striae fine and indistinctly punctured; seventh elytral interval with distinct punctures. In deciduous woodlands; v-ix. Widespread throughout Britain (except the extreme north-west) and Ireland; abundant.

Similar species: Much larger than the only other related species with four-spotted elytra, *Calodromius spilotus* and *Philorhizus quadrisignatus*.

6. *CALODROMIUS* Reitter

A small group of mainly Mediterranean arboreal species, formerly included in *Dromius*. Only one occurs in Britain and Ireland. Distinguished from related genera by the completely bordered elytral bases (Fig. 317).

1. *Calodromius spilotus* (Illiger) Plate 140

Length 3.8-4.5 mm. Head black, pronotum dark red-brown, elytra dark brown with two pale marks on each; the posterior marks are small, leaving the outer parts of the apical margin dark. Appendages yellow-brown. Pronotum small and quadrate, sides slightly sinuate and raised sharply in front of the protruding, sharp hind angles. Elytra wide relative to pronotum, basal border reaching the scutellum; striae very fine. Wings present. Mainly in coniferous or mixed woodlands; v-viii. Widespread in England, more local in Wales, Scotland and Ireland; abundant.
Similar species: Much smaller than *Dromius quadrimaculatus* which is also four-spotted. Distinguished from *Philorhizus quadrisignatus* by the protruding pronotal hind angles and complete basal border on the elytra.

7. *PHILORHIZUS* Hope

More than 35 species of small, soft-bodied mostly ground-living carabids with pale appendages, previously included in *Dromius*. The basal border of the elytra does not reach the scutellum (Fig. 318). Five species occur in Britain, two in Ireland.

Key to species of *Philorhizus*

1. Base and shoulders of elytra dark brown, in contrast to paler spots behind this .... 3. *quadrisignatus* (Dejean)

- Elytral shoulders (and often entire base) pale straw colour, sometimes with darker markings behind 2

2. Elytra entirely pale except for the suture and scutellary regions (Plate 141) .... 1. *melanocephalus* (Dejean)

- Elytra with more extensive dark markings, some of which reach the side margins 3

3. Dark mark on elytra not reaching the apex along the side margin, so that entire apex of elytra is pale (Fig. 327) .... 4. *sigma* (Rossi)

- Dark mark on elytra reaching the apex along the side margin (Figs 328, 329) 4

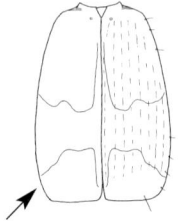

327

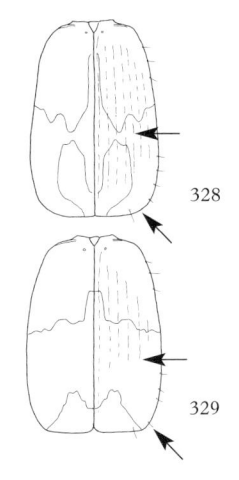

4. Pronotum darkened at least medially, notably darker than base of elytra; sides of elytra straighter, dark elytral mark more irregular, often divided into median and lateral marks (Fig. 328) ....
.... 2. *notatus* (Stephens)

- Pronotum usually entirely pale, hardly darker than elytral base; sides of elytra rounded, dark elytral mark more evenly wide, not divided (Fig. 329) .... 5. *vectensis* (Rye)

1. *Philorhizus melanocephalus* (Dejean) Plate 141

Length 2.8-3.5 mm. Head black, pronotum bright orange-red, elytra pale straw-coloured with darkened suture, as well as a triangular mark around the scutellum in front of the folded wings which show through the pale elytra. Appendages pale. Pronotum transverse, contracted and slightly sinuate in front of the sharp hind angles. Elytral sides straight, hardly widened apically. Wings present. In dry grasslands and dunes; iv-ix. Widespread in England and the east of Ireland, usually coastal in the west; very local in Scotland; very abundant.

Similar species: With a paler pronotum than *P. notatus* (2); distinguished from that species as well as the rarer *P. sigma* (4) and *P. vectensis* (5) by the lack of any dark markings on the apical third of the elytra.

2. *Philorhizus notatus* (Stephens)

Length 3-3.7 mm. Head black, pronotum dark reddish brown to almost black, elytra pale straw-coloured with darker markings along suture and apically at sides (Fig. 328); rarely these are joined to form a transverse band. Underside of abdomen black or dark brown. Appendages pale. Pronotum slightly sinuate in front of the sharp hind angles. Elytra widened apically, sides hardly rounded, striae fine. Wings usually absent. In dunes and dry grasslands; iv-viii. Widespread around the coasts of both Britain and Ireland, rarely inland; abundant.

Similar species: Pronotum generally darker than in the other species of this genus. The pale elytral shoulders distinguish it from *P. quadrisignatus* (3) and *Calodromius spilotus*. The dark lateral elytral markings are often separated from the sutural line by a pale gap, unlike the complete dark band of *P. sigma* (4) and *P. vectensis* (5).

3. *Philorhizus quadrisignatus* (Dejean)

Length 3.5-4 mm. Head black, pronotum mid reddish brown, elytra brown with a large pale mark behind each shoulder, and pale apices. Appendages pale. Head strongly constricted behind the eyes. Pronotum transverse, hardly sinuate in front of obtuse hind angles. Elytra wide and flat, striae almost absent. Wings present. In deciduous woods and shrubby land, both on trees and in litter; v-vii. Very local in eastern England, with a few records in Cornwall, south Wales and eastern Scotland; very scarce.

Similar species: The strongly constricted neck and dark elytral base distinguish this species from all other members of the genus. See also *Calodromius spilotus*.

4. *Philorhizus sigma* (Rossi)

Length 3.4-4.4 mm. Head black, pronotum bright reddish, elytra straw-coloured with a dark sutural mark and transverse band leaving apices completely pale (Fig. 327). Underside of abdomen hardly darkened. Pronotum transverse, rather contracted basally. Elytra elongate, widened apically and with shoulders completely rounded; striae very fine; Wings usually absent. In marshes, fens and at the margins of standing fresh water; iv-vi. Very local in Norfolk, Yorkshire and Lincolnshire; very scarce.
Similar species: The combination of pale pronotum and single transverse dark elytral band separate this species from all except *P. vectensis* (5). Distinguished from that species by the more elongate elytra, completely pale elytral apices and paler abdominal underside.

5. *Philorhizus vectensis* (Rye)

Length 3.2-3.7 mm. Head black, pronotum bright reddish, elytra straw-coloured with a dark transverse band sub-apically; this extends forwards medially along the suture, and backwards laterally to include the apical angles of the elytra (Fig. 329). Underside of abdomen black or dark brown. Appendages pale. Pronotum transverse, hind angles obtuse. Elytra with shoulders rounded, sides also rounded, widest just behind middle; striae distinct, with fine punctures. Wings absent. In partly vegetated dry sand or shingle near the coast; v-viii. Local along the English south coast from Cornwall to Kent, occasionally inland, Essex and Suffolk; very scarce.
Similar species: See *P. sigma* (4).

8. *MICROLESTES* Schmidt-Göbel

A large genus of very small carabids, found in all parts of the world, often only identifiable by the male genitalia. Usually smaller than *Syntomus* species, and distinguished from them by the straight, transversely truncate elytral apices (Fig. 315) and pubescent third antennal segment. Both this genus and *Syntomus* have a distinct pore at the base of the scutellary stria on each elytron. Until recently only one of the 20 European species was found in Britain but a second has occurred since 1975. Neither are found in Ireland.

Key to species of *Microlestes*

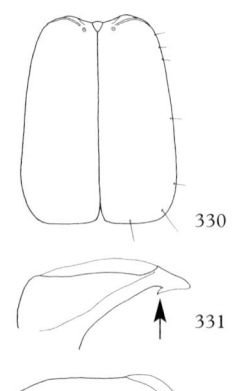

1. Length at most 2.8 mm; elytra shorter, sides rounded and diverging (Fig. 330); aedeagus very short with subapical hook (side view Fig. 331) .... 1. *maurus* (Sturm)

- Length at least 2.8 mm; elytra longer, almost parallel-sided (Plate 142); aedeagus elongate, apex not hooked (side view Fig. 332) .... 2. *minutulus* (Goeze)

1. *Microlestes maurus* (Sturm)

Length 2.3-2.8 mm. Black, shining with very indistinct microsculpture. Appendages black. Head small with moderately protruding eyes. Pronotum moderately contracted behind. Elytra with sides rather rounded (Fig. 330), widest near apex, striae almost invisible. Aedeagus short, with sub-apical ventral hook (Fig. 331) and internal teeth. Wings absent. On dry, often sandy soils with litter or shade; v-ix. Widespread in south and east England, local and coastal in south-west England and Wales; abundant.
Similar species: Smaller and often shinier than *M. minutulus* (2) with shorter, more rounded elytra.

2. *Microlestes minutulus* (Goeze) Plate 142

Length 2.8-3.7 mm. Black, moderately shining but microsculpture distinct, especially on the elytra. Appendages black. Head large with protruding eyes. Pronotum rather strongly contracted behind. Elytral sides almost straight, parallel or slightly widened behind, striae very fine. Aedeagus elongate, apex without ventral hook (Fig. 332) and internal sac without teeth. Wings present. On sandy and gravelly soils, often in open situations; v-viii. Originally restricted to the coasts of East Anglia and the Thames estuary but now widespread in much of south-east England; can be abundant.
Similar species: See *M. maurus* (1). Of similar size to *Syntomus truncatellus* but less convex with more parallel-sided elytra that are transversely, rather than obliquely, truncate apically.

9. LIONYCHUS Wissmann

There are five European species, mostly Mediterranean, in this very distinctive genus, with characteristic pronotum which is strongly contracted and almost triangular, elytral punctures (Fig. 313) and (usually) markings. One species occurs in Britain but not in Ireland.

1. *Lionychus quadrillum* (Duftschmid) Plate 143

Length 3-4 mm. Shining black, elytra typically with four pale spots, sometimes reduced to two shoulder spots only or even totally dark. Appendages black. Head flat between the eyes, finely wrinkled. Pronotum strongly contracted behind and almost triangular, with the hind angles reduced to small teeth set in front of the hind margin. Elytra rather short and wide, with fine striae separated by a row of fine punctures along each interval; apices obliquely truncate but not sinuate. Wings present. In open shingle and sand, usually near water; v-viii. Very local around the coast from Yorkshire to mid Wales, as well as inland on a few rivers in Wales and eastern England; scarce.
Similar species: Resembles a *Bembidion*, especially if the truncate elytra overlap the end of the abdomen but the apical palpal segment is not reduced as in the Bembidiini.

10. SYNTOMUS Hope

An extensive and world-wide genus, with more than 15 European species. They are separated from *Dromius* by the darkened antennae, and from *Microlestes* by the sinuate and obliquely truncate elytral apices (Fig. 316) and glabrous third antennal segment. There are three British species, two of which occur in Ireland.

Key to species of *Syntomus*

1. Metallic bronze, elytra each with three deep foveae (Plate 144) .... 1. *foveatus* (Geoffroy in Fourcroy)

- Non-metallic black, elytra without foveae 2

2. Elytra more elongate and parallel-sided, with pale shoulder mark (Fig. 333) .... 2. *obscuroguttatus* (Duftschmid)

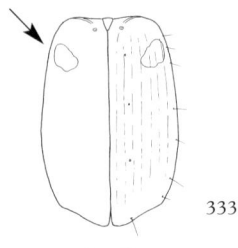

- Elytra shorter, more rounded (Fig. 334) and uniformly black .... 3. *truncatellus* (Linnaeus)

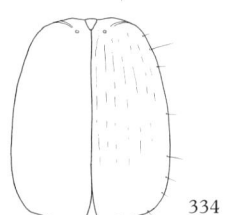

1. *Syntomus foveatus* (Geoffroy in Fourcroy) Plate 144

Length 3.2-3.8 mm. Uniformly dark metallic bronze, strongly microsculptured. Appendages black. Eyes large and prominent. Pronotal sides almost straight, contracted behind. Elytra elongate with sides only slightly rounded, widest behind middle; base almost completely bordered; striae punctured but faint, third interval with two deeply foveate dorsal punctures. Wings absent. On dry heaths, waste ground, grasslands, arable land and dunes; iii-ix. Widespread in much of eastern England, and on the west coasts of both England and Wales as well as the south-east of Ireland; local in Scotland except the north; very abundant.

2. *Syntomus obscuroguttatus* (Duftschmid)

Length 3-3.8 mm. Black, or with slightly paler elytra, which usually have an indistinct pale shoulder mark. Antennae black; legs brown with darker femora. Pronotal sides slightly rounded in front of very obtuse hind angles. Elytra elongate with distinct fine microsculpture, almost parallel-sided (Fig. 333); striae fine but distinct, dorsal punctures small and not foveate. Wings absent. In litter and moss, usually in damp situations or on clay soils; iii-x. Widespread in southern England, not extending north beyond Lincolnshire; scarce.
Similar species: Without the foveate dorsal punctures and metallic colour of *S. foveatus* (1) more elongate than *S. truncatellus* (3) which lacks a pale shoulder spot on the elytra.

3. *Syntomus truncatellus* (Linnaeus)

Length 2.8-3.3 mm. Black, moderately shining with very fine microsculpture. Antennae black; legs brown with darker femora. Pronotum very transverse, sides slightly rounded in front of almost rounded hind angles. Elytra short and rather convex, sides rounded (Fig. 334); striae barely distinct, dorsal punctures small and not foveate. Wings usually absent. On open ground in fields, pasture woodland and dunes; iv-viii. Very local in eastern

England as well as occasionally on the coasts of south-west England, Wales, Scotland and south-east Ireland; scarce.
Similar species: See *S. obscuroguttatus* (2) and *Microlestes minutulus*.

Tribe ODACANTHINI

This tribe has only two genera in Europe, each with a single distinctive species. Only one of these, *Odacantha* Paykull, is found in northern Europe including Britain but not Ireland. Characterised by the truncate elytra, cylindrical pronotum and first antennal segment not elongated (cf. *Drypta*).

1. *Odacantha melanura* (Linnaeus) Plate 145

Length 6.5-7.5 mm. Head and pronotum black with a green or blue metallic reflection, elytra reddish brown with a common black apical spot. Antennae black with three pale basal segments; legs brown with apices of femora and most of the tarsi black. Eyes protruding, head gradually contracted behind, forming a neck in front of the very narrow, cylindrical pronotum. Pronotum with strong transverse striations. Elytral striae comprising fine rows of punctures; apical margin slightly sinuate, with about four setae. Wings present. In and around reedbeds; iv-vii. Local and often coastal in southern England, East Anglia and south Wales; can be abundant.

Tribe DRYPTINI

There are about 50 species in the very distinctive, mostly tropical genus *Drypta* Latreille, with cylindrical pronotum and characteristic antennae (Fig. 44). One of the two European species occurs in Britain but not in Ireland.

1. *Drypta dentata* (Rossi) Plate 146

Length 7-8.5 mm. Shining blue or green metallic, dorsal surface strongly punctured and with pale pubescence. Appendages brown, apex of first antennal segment black; tarsi and third antennal segment sometimes also darkened. First antennal segment longer than head, swollen apically, second segment extremely small. Tarsi with fourth segment bilobed. Mandibles elongate and protruding. Eyes large and almost hemispherical. Pronotum elongate, narrower than head, cylindrical without defined side margins. Elytra broad, widening apically with strongly punctured striae; apical margin slightly sinuate. Wings present. On coastal, partly vegetated sandy cliffs and slippages near fresh water; v-vi. Extremely local in Dorset and the Isle of Wight; very scarce.

Tribe ZUPHIINI

A small, mainly tropical, group with about ten European species in three genera. There is a single British species in the genus *Polistichus* Bonelli. It is flatter than *Cymindis*, with head and pronotum more elongate, antennae entirely pubescent and elytra with a fine membranous apical border (Fig. 43).

1. *Polistichus connexus* (Fourcroy) Plate 147

Length 8-9.5 mm. Dorsal surface flat and pubescent throughout; head and pronotum red-brown, elytra dark brown-black with an elongate pale area covering most of intervals

3-6 except the extreme apices. Appendages pale brown. Antennae entirely pubescent, basal segment wider than the remainder. Tarsal claws not toothed. Head sparsely but very strongly punctured; strongly contracted behind the protruding eyes, which have a strong row of curved hairs behind. Pronotum densely punctured, not transverse, hind angles sharp but slightly obtuse. Elytra densely punctured, intervals flat; apices transversely truncate. Wings present. In cracks and crevices, usually in clay soils or cliff bases but also on sandy or gravelly soils; v-viii. Very local and usually coastal in the extreme south and east of England, can be abundant.

References

Anderson, R. 1985. *Agonum lugens* (Duftschmid) new to the British Isles (Col., Carabidae). *Entomologist's Monthly Magazine* **121**: 133-135.

Anderson, R. & Luff, M.L. 1994. *Calathus cinctus* Motschulsky, a species of the *Calathus melanocephalus / mollis* complex (Col., Carabidae) in the British Isles. *Entomologist's Monthly Magazine* **130**: 131-135.

Anderson, R., McFerran, D. & Cameron, A. 2000. *The ground beetles of Northern Ireland (Coleoptera-Carabidae)*. Ulster Museum, Belfast.

Anderson, R., Nash, R. & O'Connor, J.P. 1997. Irish Coleoptera. A revised and annotated list. *Irish Naturalists' Journal, Special Entomological Supplement*: 1-81.

Arndt, E. 1991. Familie Carabidae. In Klausnitzer, B.: *Die Larvern der Käfer Mitteleuropas. 1. Adephaga*. Goecke & Evers, Krefeld: pp 45-141.

Arndt, E. 1998. Phylogenetic investigation of Carabidae (Coleoptera) using larval characters. In: Ball, G.E., Casale, A. & Tagliante, A.V. (eds.): *Phylogeny and classification of Caraboidea. (Coleoptera: Adephaga)*. Regional Science Museum, Turin: pp 171-190.

Ball, G.E., Casale, A. & Tagliante, A.V. 1998. Introduction. In: Ball, G.E., Casale, A. & Tagliante, A.V. (eds.): *Phylogeny and classification of Caraboidea. (Coleoptera: Adephaga)*. Regional Science Museum, Turin: pp 15-52.

Casale, A. & Bfiezina, B. 2003. Checklist. In: Turin, H., Penev, L. & Casale, A. (eds.), *The genus Carabus in Europe. A synthesis*. Pensoft, Sofia/Moscow: pp 15-71.

Coulon, J. 2006. Révision des taxons d'Europe et du basin méditerranéen occidental rattaché à *Bembidion (Peryphus) cruciatum* Dejean. (Coleoptera, Carabidae, Bembidiini). *Nouvelle Revue d'Entomologie* **22** (2005): 327-350.

Crossley, R. & Norris, A. 1975. *Bembidion humerale* Sturm (Col., Carabidae) new to Britain. *Entomologist's Monthly Magazine* **111**: 59-60.

Deuve, T. 1993. L'abdomen et les genitalia des femelles de Coléoptères Adephaga. *Mémoires du Muséum national d'Histoire naturelle, Zoologie* **155**: 1-185.

Deuve, T. 1994. *Une classification du genre Carabus*. Sciences Nat, Venette.

Erwin, T.L. & Sims, L.L. 1984. Carabid beetles of the West Indies (Insecta: Coleoptera). A synopsis of the genera and checklist of tribes of Caraboidea and of the West Indian species. *Quaestiones entomologicae* **20**: 350-466.

Erwin, T.L. 1985. The taxon pulse: a general pattern of lineage radiation and extinction among carabid beetles. In: Ball, G.E. (ed.): *Taxonomy, phylogeny and zoogeography of beetles and ants*. Junk, Dordrecht: pp 437-493.

Eversham, B. & Collier, M. 1997. *Microlestes minutulus* (Goeze) (Carabidae) new to Britain. *The Coleopterist* **5**: 93-94.

Forel, J. & Leplat, J. 2001, 2003, 2005. *Faune des carabiques de France – I, XI, X*. Magellanes, Andrésy.

Forsythe, T.G. 2000. *Ground beetles. Naturalists' Handbooks* **8** (2nd edn.). Richmond Publishing, Slough.

Hammond, P.M. 1982. *Cymindis macularis* (Fischer v. Waldheim) (Col., Carabidae) – apparently a British species. *Entomologist's Monthly Magazine* **118**: 37-38.

Hansen, M. 1996. Katalog over Danmarks biller. *Entomologiske Meddelelser* **64** (1-2): 1-231.

Hůrka, K. 1996. *Carabidae of the Czech and Slovak Republics*. Kabourek, Zlin.

Kryzhanovskij, O.L., Belousov, I.A., Kabak, I.I., Kataev, B.M., Makarov, K.V. & Shilenkov, V.G. 1995. *A checklist of the ground-beetles of Russia and adjacent lands (Insecta, Coleoptera, Carabidae)*. Pensoft, Sofia.

Lawrence, J.F. & Newton, A.F. 1995. Families and subfamilies of Coleoptera (with selected genera, notes, references and data on family names). In: Pakaluk, J. & Slipinski, S.A. (eds.): *Biology, phylogeny and classification of Coleoptera*. Museum of the zoological Institute PAN, Warsaw: pp 779-1006.

Levey, B. & Pavett, P.M. 1999. *Bembidion (Pseudolimnaeum) inustum* Duval, (Coleoptera: Carabidae) an interesting new addition to the British fauna. *British Journal of Entomology & Natural History* **11**: 169-171.

Lindroth, C.H. 1974. Coleoptera Carabidae. *Handbooks for the Identification of British Insects.* **4** (2): 1-148.

Lindroth, C.H. 1985-1986. *The Carabidae (Coleoptera) of Fennoscandia and Denmark. Fauna entomologica Scandinavica* **15**(1-2), Brill, Leiden.

Löbl, I & Smetana, A. (eds.) 2003. *Catalogue of Palaearctic Coleoptera, Vol. 1: Archostemata, Myxophaga, Adephaga.* Apollo Books, Strenstrup.

Lohse, G.A. & Lucht, W.H. 1989. *Die Käfer Mitteleuropas. 1 Supplementband mit Katalogteil.* Goecke & Evers, Krefeld.

Luff, M.L. 1990. *Pterostichus rhaeticus* Heer (Col., Carabidae), a British species previously confused with *P. nigrita* (Paykull). *Entomologist's Monthly Magazine* **126**: 245-249.

Luff, M.L. 1993. *The Carabidae (Coleoptera) larvae of Fennoscandia and Denmark. Fauna entomologica Scandinavica* **27**, Brill, Leiden.

Luff, M.L. 1998. *Provisional atlas of the ground beetles (Coleoptera, Carabidae) of Britain.* Biological Records Centre, Huntingdon.

Luff, M.L. 2005. Biology and ecology of immature stages of ground beetles (Carabidae). In: Lövei, G. & Toft. S. (eds.) *European Carabidology 2003. Proceedings 11th European carabidologists' meeting.* Danish Institute of Agricultural Sciences Report **114**: pp 183-208.

Müller-Motzfeld, G. (ed.) 2004. Bd. 2 *Adephaga 1: Carabidae (Laufkäfer).* In Freude, H., Harde, K.W., Lohse, G.A. & Klausnitzer, B., *Die Kafer Mittteleuropas.* Spektrum-Verlag, Heidelberg/Berlin, 2nd edition.

Owen, J.A. 1996. *Harpalus (Pseudophonus) griseus* Panzer (Col.: Carabidae) at Wimbledon, Surrey – the first definitely British record? *Entomologist's Record & Journal of Variation* **108**: 69-72.

Pope, R.D. 1977. A check list of British Insects, 2nd edition. Part 3: Coleoptera & Strepsiptera. *Handbooks for the Identification of British Insects.* **11** (3): 1-105.

Roig-Juñent, S. 1998. Cladistic relationships of the tribe Broscini (Coleoptera: Carabidae). In: Ball, G.E., Casale, A. & Tagliante, A.V. (eds.): *Phylogeny and classification of Caraboidea. (Coleoptera: Adephaga).* Regional Science Museum, Turin: pp 343-358.

Sciaky, R. 1987. Revisione delle specie paleartiche occidentali del genere *Ophonus* Dejean, 1821. *Memorie delle Società Entomologica Italiana* **65** (1986): 29-120.

Serrano, J. & Ortuño, V.M. 2001. Revisión de las especies ibéricas de *Bradycellus* Erichson (Coleoptera, Carabidae, Harpalini). *Bulletin de la Société entomologique de France* **106**: 337-348.

Speight, M.C.D., Martinez, M. & Luff, M.L. 1986. The *Asaphidion* (Col.: Carabidae) species occurring in Great Britain and Ireland. *Proceedings & Transactions of the British Entomological & Natural History Society* **19**: 17-21.

Telfer, M.G. 2001a. *Bembidion coeruleum* Serville (Carabidae) new to Britain and other notable carabid records from Dungeness, Kent. *The Coleopterist* **10**: 1-4.

Telfer, M.G. 2001b. *Ophonus subsinuatus* Rey (Carabidae) new to Britain, with a discussion of its status. *The Coleopterist* **10**: 39-43.

Telfer, M.G. 2003. *Acupalpus maculatus* Schaum, 1860: another carabid new to Britain from Dungeness. *The Coleopterist* **12**: 1-6.

Thiele, H.-U. 1977. *Carabid beetles in their environment.* Springer-Verlag, Berlin.

Turin, H. 1990. Naamlijst voor de Nederlandse loopkevers (Coleoptera: Carabidae). *Entomologische Berichten, Amsterdam* **50**: 61-72.

Turin, H. 2000. *De Nederlandse Loopkevers. Verspreiding en Oecologie [The Netherlands Ground Beetles. Distribution and Ecology].* KNNV Uitgeverij.

Turin, H., Casale, A., Kryzhanovskij, O.L., Makarov, K.V. & Penev, L.D. 1993. Checklist and atlas of the genus *Carabus* Linnaeus in Europe (Coleoptera, Carabidae). Backhuys, Leiden.

Turin, H., Penev, L. & Casale, A. (eds.) 2003. *The genus Carabus in Europe. A synthesis. Fauna Europaea Evertebrata* **2**, Pensoft, Sofia/Moscow.

Index

Main entries, and start of sections for genus, subgenus and species are shown in **bold.** Synonyms are given in italic.

ABAX 5, 19, 105, **115**
ACTEDIUM 15, 77, **82**
acuminata, Amara 22
ACUPALPUS 26, 148, 149, 172, **173**, 177
adamantina, Amara 23
ADELOSIA 19, **112**
adstrictus, Pterostichus 19, 109, **113**
adustum, Bembidion 16
aenea, Amara 22, 135, **138**-140, 142
aeneum, Bembidion 18, 100, **101**
aeneus, Dyschirius 13, 59, **60**-62
aeneus, Dyschirus 13
aeneus, Harpalus 23
AEPOPSIS 14
AEPUS 2, 14, **63**
Aepus marinus 14, **64**, 65
Aepus robinii 14, 64, **65**
aesthuans, Notiophilus 12, 51, **52**
aestuans, Notiophilus 12
aethiops, Pterostichus 19, **111**, 112
AETOPHORUS 28
affine, Bembidion 17
affinis, Harpalus 1, 23, 151, **153**, 155, 156, 166
affinis, Poecilus 18
afrum, Agonum 22
agilis, Dromius 28, 192, **193**
AGONODERUS 26
AGONUM 2, 4, 7, 8, 21, 22, 116, 118, 125, **126, 129**
albipes, Paranchus 21, **124**
alpinus, Curtonotus 23, **147**
AMARA 3, 7, 8, 22, 23, 133, **134, 136**, 157, 180
ambiguus, Calathus 20, 119, **121**
AMPHIGYNUS 20, **120**
ANCHOMENUS 21, 117, **124**
ANCHUS 21
andreae, Bembidion 16
anglicanus, Carabus 10
anglicus, Stenolophus 26
angustatus, Dyschirius 13, 58, **59**-61

angustatus, Pterostichus 19
angusticollis, Ophonus 25
angusticollis, Platynus 21
angustus, Dromius 28, 192, **193**
ANISODACTYLUS 25, 149, **165**
anomalus, Badister 27
anthobia, Amara 22, 136, **139**-141
anthracinus, Pterostichus 19, 110, **114**, 115
ANTHRACUS 26, 148, 149, **177**
antiquorum, Bembidion 17
anxius, Harpalus 23, 153, **154**, 156, 157
apricaria, Amara 23, **145**, 146
aquaticus, Notiophilus 12, 51, **52**
ARCHICARABUS 11, **43**
arctica, Miscodera 13, **62**
ardosiacus, Ophonus 24, 159, **161**, 162
arenosus, Dyschirus 13
areolatus, Perileptus 14, **64**
argenteolum, Bracteon 15, **76**
ARGUTOR 19, **115**
articulatum, Bembidion 17, 98, **99**
arvensis ssp. *sylvaticus, Carabus* 10
arvensis, Carabus 8, 10, 40, **41**, 42, 43, 44
ASAPHIDION 15, 69, **73**
assimile, Bembidion 17, 95, **96**
assimilis, Patrobus 18, **104**
assimilis, Platynus 21, **125**
ater, Abax 19
aterrimus, Pterostichus 19, 108, **112**
atratum, Agonum 22
atratus, Syntomus 29
atricapillus, Demetrias 28, 188, **189**
atricornis, Anisodactylus 25
atrocaeruleum, Bembidion 16, 85, **86**, 87
atrocoeruleum, Bembidion 16
atrorufus, Patrobus 18, **104**
atroviolaceum, Bembidion 17
attenuatus, Harpalus 24, 152, **154**, 156
aulicus, Curtonotus 23, **147**, 148
auratus, Carabus 11, 40, **41**, 43

203

AUTOCARABUS 11, **41**
axillaris, Cymindis 28, **190**, 191
azureus, Ophonus 24, 159, **162**-164

BADISTER 1, 4, 27, 180, **181**, 182
Badistriini 7
Badistrini 180
balbii, Nebria 12
basalis, Cymindis 28
BATENUS 21, 118, **125**
BAUDIA 27, 181, **183**, 184
Bembidiini 14, 35, 69, 197
BEMBIDION 1, 2, 5, 7, 8, 15, 17, 69, 70, 74, 75, **76**, 77, 93, **97**, 173, 197
BEMBIDIONETOLITZKYA 16, 79, 85
bifrons, Amara 23, **143**, 145
biguttatum, Bembidion 18, 100, **101**, 102
biguttatus, Notiophilus 12, 51, **52**-54
bimaculatus, Dromius 28
binotatus, Anisodactylus 25, 165, **166**
bipunctatum, Bembidion 15, **82**
bipunctatus, Lionychus 29
bipustulatus, Badister 27
bipustulatus, Panagaeus 27, **184**, 185
bistriatus, Tachys 14, **71**, 72
bisulcatus, Porotachys 9
blacki, Notiophilus 12
BLECHRUS 29
BLEMUS 14, 63, **68**
BLEPHAROPLATAPHUS 16
BLETHISA 12, **54**
borealis, Pelophila 12, **50**, 54
BOREONEBRIA 12, **49**
BOTHRIOPTERUS 19, **113**
Brachininae 10, 30
BRACHINUS 4, 5, 10, **32**, 33
BRACHYNIDIUS 10, **33**
BRACTEON 8, 15, 70, **75**, 76
BRADYCELLUS 2, 8, 25, 26, 149, **169**, 172, 173
BRADYTUS 23, 134, **145**
brevicollis, Nebria 5, 12, 48, **49**
brevicollis, Ophonus 25
britannicus, Carabus 11
Broscidae 7
Broscinae 7

Broscini 3, 13, 35, 62
BROSCUS 13, **62**
browni, Carabus 11
brunneipes, Acupalpus 26
brunnipes, Acupalpus 26, 174, **175**, 176
bruxellense, Bembidion 16, **89**, 90-93
bualei ssp. *anglicanum, Bembidion* 16
bualei ssp. *polonicum, Bembidion* 16
bualei, Bembidion 16, 89, **90**-93
bullatus, Badister 27, 181, **182**, 183

caeruleipenne, Bembidion 15
caerulescens, Poecilus 18
caeruleum, Bembidion 8, 16, 85, **86**, 87
caesica, Poecilus 18
CALATHUS 20, 116, **118**, **121**, 119
calceatus, Harpalus 23, 149, 150
CALLISTUS 1, 27, 177, **179**
callosum, Bembidion 9
CALODROMIUS 28, 187, 192, **194**
CALOSOMA 4, 10, 38, **39**
campestris, Cicindela 10, **31**
cancellatus, Carabus 9
Carabinae 7, 10, 30, 33
Carabini 4, 10, 34, 38
caraboides ssp. *rostratus, Cychrus* 11
caraboides, Cychrus 11, **44**
CARABUS 3, 5, 7, 8, 10, 38, **39**, **42**
caspius, Harpalus 24
catenulatus, Carabus 11
caucasicus, Bradycellus 26, **170**-172
celere, Bembidion 15
CELIA 23, 134, **143**-145
cephalotes, Broscus 13, **62**
CHAETOCARABUS 11, **42**
chalceus, Pogonus 18, 102, **103**
chalybeum, Agonum 22
championi, Ophonus 25
CHLAENIELLUS 26
Chlaeniini 26, 37, 177
CHLAENIUS 1, 26, 177, **178**
chlorocephala, Lebia 28, 187, **188**
chrysocephala, Lebia 28
CICINDELA 1, 10, 30, **31**
Cicindelinae 10, 12, 30

CILLENUS 8, 15, 70, **75**
cinctus, Calathus 8, 20, 118, 120, **121**, 122
cisteloides, Calathus 20
clarki, Bembidion 17
clarkii, Bembidion 17, 95, **96**
clathratus, Carabus 10
clatratus, Carabus 10, 40, **42**
clavipes, Patrobus 18
CLIVINA 1, 4, 13, **57**, 58
coeruleotinctum, Bembidion 15
coerulescens, Loricera 13
coeruleum, Bembidion 16
cognatus, Trichocellus 26, **168**, 169
collaris, Badister 27, 182, **183**, 184
collaris, Bradycellus 26
collaris, Clivina 13, **57**
COLLIURIS 29
comma, Stenolophus 9
communis, Amara 22, 138, **139**, 141
complanata, Amara 23
complanata, Eurynebria 12, **50**
complanatus, Laemostenus 20, **123**
concinnum, Bembidion 16
concinnus, Pterostichus 19
connexus, Polistichus 29, **199**
consentaneus, Harpalus 24
consitus, Carabus 11
consputus, Anthracus 26, **177**
consularis, Amara 23, 144-**146**
continua, Amara 22
contracta, Clivina 13
convexior, Amara 22, 138, **139**, 141
convexiusculus, Curtonotus 23, **147**
convexus, Carabus 9
cordatus, Ophonus 25, 159, **162**-165
crenatus, Pterostichus 19
crepitans, Brachinus 10, 32, **33**
cristatus ssp. *parumpunctatus, Pterostichus* 19
cristatus, Pterostichus 19, 107, 108, **111**-113
cruciatum, Bembidion 16
cruxmajor, Panagaeus 27, 184, **185**
cruxminor, Lebia 28, 187, **188**
CRYPTOPHONUS 24, 150, **158**
csikii, Bradycellus 26, 170, **171**
cupreus, Elaphrus 12, **55**, 56

cupreus, Harpalus 24, 149, 152
cupreus, Poecilus 18, **106**
cursitans, Amara 9, 135, 143
curta, Amara 22, 136, **139**, 141, 142
CURTONOTUS 8, 23, 133, 134, **146**, 147
curtum, Asaphidion 8, 15, **73**, 74
cyaneotinctum, Bembidion 15
cyanocephala, Lebia 28, 187, **188**
Cychrini 11, 34, 44
CYCHRUS 2-4, 11, **44**
CYLINDERA 10, 30, **32**
CYMINDIS 28, 186, **190** 199
CYRTONOTUS 23

dahli, Agonum 22
dalmatinum var. *latinum, Bembidion* 16
DANIELA 16
decorum, Bembidion 16, 87, **90**, 92, 93
degenerata, Nebria 12
deletum, Bembidion 16, 88, 89, **90**, 92
DEMETRIAS 28, 186, **188**, **189**, 191
Demetriini 186
dentata, Drypta 29, **199**
dentellum, Bembidion 16, **82**, 84
depressus, Licinus 27, **180**
depressus, Platyderus 20, **118**
derelictus, Acupalpus 26
DIACHROMUS 25, 148, **167**
DICHEIROTRICHUS 25, 148, **167**, 168, 172
DICHIROTRICHUS 25
diffinis, Ophonus 24
dilatatus, Badister 27, 182, **183**, 184
diligens, Pterostichus 19, 110, 114, **115**
dimidiatus, Harpalus 24, 141, **154**, 156
dimidiatus, Poecilus 18
dinniki, Poecilus 18
DIPLOCAMPA 17, 78, **95**
discoideus, Harpalus 24
discus, Blemus 14, **69**, 177
discus, Dromius 28
distinctus, Bradycellus 26, 170, **171**, 172
doris, Bembidion 17, **98**, 99
dorsalis, Acupalpus 26
dorsalis, Anchomenus 21, **124**
dorsuarium, Bembidion 16

Dromiini 186
DROMIOLUS 29
DROMIUS 1, 4, 5, 28, 187, 188, 191, **192**, 194, 197
DRYPTA 3, 29, **199**
Dryptini 29, 38, 199
dubius, Acupalpus 26, 174, **175**, 176
DYSCHIRIODES 13, **60**
DYSCHIRIUS 1, 2, 13, **57**, 58, **59**

edmondsi, Tachys 14
Elaphrini 12, 34, 54
ELAPHROPUS 15, 70-**72**, 75
ELAPHRUS 2, 12, **54**, **55**
elegans, Acupalpus 26, 174, **175**-177
elongatulus, Dyschirus 13
elongatum, Agonum 22
emarginata, Drypta 29
emarginatum, Agonum 22, **130**-133
EMPHANES 17, 79, **96**
EPAPHIUS 14, **66**
ephippium, Bembidion 16, **84**
equestris, Amara 23, 144, **146**
ericeti, Agonum 22, 129-**131**, 132
erratus, Calathus 20, 119, **121**
erythrocephalus, Harpalus 24
erythroderus, Calathus 20
erythropus, Poecilus 18
erythropus, Pterostichus 19
EUCARABUS 8, 10, **41**
EUPETEDROMUS 16, 79, **82**, 83
EUROPHILUS 8, 21, **126**, 129
EURYNEBRIA 12, 45, **50**
eurynota, Amara 22, 137, 138, **140**, 142
exasperatus, Carabus 11
excavatus, Patrobus 18
exiguus, Acupalpus 26, 174-**176**
extensus, Dyschirius 13, 58, **59**

famelica, Amara 22, 136, 138, **140**-142
familiaris, Amara 22, 136, 139, **140**, 141
fasciolatus, Polistichus 29
femoratum, Bembidion 16, 89, 90, **91**-93
FERONIA 18
ferrugineus, Leistus 11, 46, **47**
flammulatum, Bembidion 16

flavicollis, Acupalpus 26, 174-**176**, 177
flavipes, Asaphidion 15, 73, **74**
flavipes, Calathus 20
fluviatile, Bembidion 16, 88, **91**, 93
fossor, Clivina 13, **57**
foveatus, Syntomus 29, **198**
foveola, Syntomus 29
froelichi, Harpalus 24
froelichii, Harpalus 24, 152, **154**, 156, 157, 158
fuliginosum, Agonum 21, **127**, 128, 132
fulva, Amara 23, 145, **146**
fulvescens, Aepus 14
fulvibarbis, Leistus 11, 46, **47**
fulvipes, Calathus 20
fulvus, Trechus 14, 65, **67**, 68
fumigatum, Bembidion 17, 83, 95, **96**
funestus, Harpalus 24
fusca, Amara 23, **143**, 146
fuscipes, Calathus 20, 119, **121**
fuscus, Calathus 20

gallicus, Carabus 11
genei, Bembidion 17
geniculatum, Bembidion 16, **86**, 87
germanica, Cylindera 10, **32**
germanus, Diachromus 25, **167**
germinyi, Notiophilus 12, 51, **52**, 53
gibbus, Dyschirus 13
gibbus, Zabrus 22
gilvipes, Bembidion 17, **95**, 96
glabratus ssp. *lapponicus, Carabus* 11
glabratus, Carabus 11, 40, **42**, 44
glabratus, Microlestes 29
globosus, Dyschirius 13, 58, **60**
gracile, Agonum 21, **127**, 128
gracilipes, Agonum 22, 129, **131**
gracilis, Pterostichus 19, 110, **114**, 115
granulatus ssp. *granulatus, Carabus* 10
granulatus ssp. *hibernicus, Carabus* 10
granulatus, Carabus 10, 40, 41, **42**, 44
gregarius, Tachys 14
griseus, Harpalus 8, 23, **150**, 151
gustavi, Dicheirotrichus 25
gustavii, Dicheirotrichus 25, 167, **168**
guttula, Bembidion 18, 100, **101**, 102

gyllenhali, Nebria 12

haemorrhoum, Bembidion 18
HAPLOHARPALUS 23
Harpalini 4, 5, 7, 23, 37, 148
harpalinus, Bradycellus 26, 169, **171**, 172
harpaloides, Ocys 15, 74, **75**
HARPALUS 2, 4, 5, 7, 8, 23, 24, 146, 148, **149**, 150, **151**, 156, 157, 158
HELOBIUM 12
helopioides, Oodes 27, **179**
HEMICARABUS 11, **43**
holosericeus, Chlaenius 27
honestus, Harpalus 24, 152, 153, **155**, 156
humerale, Bembidion 8, 17, **97**
humeralis, Badister 27
hybrida, Cicindela 10, **31**
hypocrita, Notiophilus 12

iberica, Nebria 12
ignavus, Harpalus 24
illigeri, Bembidion 17, 87, **93**, 98
imperialis, Demetrias 28, **189**
impunctipennis, Dyschirius 13, 59-**61**
inaequalis, Pterostichus 19
infima, Amara 23, 142, 143, **144**, 156
inquisitor, Calosoma 10, **39**
insignis, Philorhizus 29
insularis, Carabus 11
intricatus, Carabus 11, 40, **42**, 44
inustum, Bembidion 8, 17, **94**
iricolor, Bembidion 18, 100, **101**, 102

junceus, Platynus 21

kineli, Badister 27
klinckowstroemi, Nebria 12
kugelanni, Poecilus 18, 106, **107**

LAEMOSTENUS 20, 116, **123**
LAEMOSTHENES 20
laevipes, Harpalus 24, 153, **155**, 157
LAGARUS 19, **115**
LAMPRIAS 28, **188**
lampros, Bembidion 15, 80, **81**, 97

lapidosus, Trechus 14
lapponicus, Elaphrus 13, **55**, 56
LASIOTRECHUS 14
lateralis, Cillenus 15, **75**
lateralis, Nebria 12
laticollis, Masoreus 28
laticollis, Ophonus 25, 159, 162, **163**
latus, Harpalus 24, 153, **155**-157
LEBIA 5, 28, 186, **187**, 188
Lebiini 6, 28, 38, 186
LEISTOPHORUS 11, **46**
LEISTUS 4, 11, **45**, 46
LEJA 17
lepidus, Poecilus 18, 106, **107**
leucophthalmus, Sphodrus 20, **122**
leucopthalmus, Sphodrus 20
Licinini 2, 4, 7, 27, 36, 180
LICINUS 27, **180**, 181
limbatum, Omophron 10, **33**
LIMNOCARABUS 10, **42**
linearis, Paradromius 28, 189, **191**, 192
LIONYCHUS 26, 186, **197**
litorale, Bracteon 15, **76**
litoralis, Pogonus 18
littorale, Bembidion 17
littorale, Bracteon 15
littoralis, Pogonus 18, 102, **103**
livens, Batenus 21, **125**
livida, Amara 23
livida, Nebria 12, **48**
longiceps, Paradromius 28, 191, **192**
longicollis, Pterostichus 19, 108, **111**, 112, 115
longicornis, Thalassophilus 14, **68**
LOPHA 17
LORICERA 2, 4, 13, **56**
Loricerini 13, 34, 56
LOROCERA 13
lucida, Amara 22, 136, 139, **140**, 142
luedersi, Dyschirius 13, 59-**61**, 62
lugens, Agonum 8, 22, 130, **131**
lunatum, Bembidion 16, 89, **91**, 93
lunatus, Callistus 27, **179**
lunicollis, Amara 22, 136, 139, 140, **141**, 142
lunulatum, Bembidion 18, 100, 101, **102**
luridipennis, Pogonus 18, 102, **103**

luridus, Acupalpus 26
luteatus, Acupalpus 26
luteicornis, Harpalus 24
LYMNAEUM 17, 80, **94**
LYPEROSOMUS 19, **112**

macer, Pterostichus 1, 19, 109, **112**
macularis, Cymindis 8, 28, **190**, 191
maculatus, Acupalpus 8, 26, 175, **176**, 177
madidus, Pterostichus 19, **111**, 112, 113
mannerheimi, Bembidion 18
mannerheimii, Bembidion 18, 95, 100, 101, **102**
marginata, Lebia 9, 187
marginatum, Agonum 22, 129, **131**
marinus, Aepus 14, **64**, 65
maritima, Cicindela 10, 31, **32**
maritimum, Bembidion 16, 89, **91**
Masoreini 28, 38, 185
MASOREUS 28, **185**
maurus, Microlestes 29, 196, **197**
MEGODONTUS 11, **44**
melanarius, Pterostichus 19, 108, 111, 112, **113**-115, 166
melancholicus, Harpalus 24, 156, **158**
melanocephalus, Calathus 20, 120, 121, **122**
melanocephalus, Philorhizus 29, 194, **195**
melanocornis, Chlaenius 26
melanura, Odacantha 29, **199**
melleti, Ophonus 25
melletii, Ophonus 25, 161, **163**-165
meridianus, Acupalpus 26, 174, **176**, 177
meridionalis, Badister 27, 181, **182**
meridionalis, Dromius 28, 192, **193**
MESOCARABUS 11, **44**
METABLETUS 29
metallescens, Harpalus 24
METALLINA 15, 78, **81**
METOPHONUS 25, 158, **162**, 163
micans, Agonum 21, 127, **128**
MICROLESTES 29, 186, **196**, 197
micropterus, Calathus 20, 120, **122**
micros, Tachys 14, **71**
micros, Trechoblemus 14, **69**, 177
minimum, Bembidion 17, 96, **97**, 100
minor, Pterostichus 19, 110, **114**, 115

minutulus, Microlestes 8, 29, 196, **197**, 199
minutus, Trechus 14
MISCODERA 13, **62**
mixtus, Stenolophus 26, 172, **173**
moestum, Agonum 22
mollis, Calathus 20, 120, 121, **122**
monilis, Carabus 11, 40, 41, **43**, 44
monostigma, Demetrias 28, **189**
montanus, Leistus 11, **46**, 47
monticola, Bembidion 16, 88, 90, **92**
monticola, Ophonus 24
montivaga, Amara 23, 137, **141**, 142
montivagus, Harpalus 24
MORPHOCARABUS 11, **43**
muelleri, Agonum 22, 129, 131, **132**
multipunctata, Blethisa 10, 50, **54**

NEBRIA 5, 12, 45, **48**, **49**, 50
Nebriini 4, 11, 34, 45
nebulosum, Bembidion 16
neglectus, Harpalus 24, 152, 154-**156**
NEJA 15, 78, **80**
nemoralis, Carabus 11, 40, **43**, 44
nemorivagus, Anisodactylus 25, 165, **166**
NEPHA 17, 77, **93**
niger, Pterostichus 19, 108, 111, **112**, 113, 122
nigriceps, Perigona 27, **185**
nigricorne, Bembidion 15, **80**, 81, 100
nigricornis, Chlaenius 26, **178**, 179
nigrita, Pterostichus 4, 19, 110, 113, **114**, 115
nigriventris, Philorhizus 29
nigropiceum, Bembidion 17, **94**
nigrum, Agonum 22, 127, 130, **132**
nitens, Carabus 11, 40, 41, **43**
nitida, Amara 23, 137, 138, **141**, 142
nitidulum, Bembidion 16
nitidulus, Chlaenius 27, **178**
nitidulus, Ophonus 25
nitidus, Dyschirius 13, 57, 59, **61**
nivalis, Nebria 12, 48, **49**
nivalis, Synuchus 20
normannum, Bembidion 17, 96, **97**
NOTAPHEMPHANES 16, 79, **84**
NOTAPHUS 16, 79, **83**
notatus, Philorhizus 29, **195**

NOTHAPHEMPHANES 16
Notiophilini 12, 34, 50
NOTIOPHILUS 2, 4, 5, 12, **50**
nubigena, Calathus 20

obliquum, Bembidion 16, **83**, 84
oblongiusculus, Scybalicus 25, **167**
oblongopunctatus, Pterostichus 19, 109, **113**
oblongus, Oxypselaphus 21
oblongus, Paranchus 21
obscuroguttatus, Syntomus 29, **198**, 199
obscurus, Dyschirius 13, 58, **60**, 61
obscurus, Ophonus 24
obscurus, Oxypselaphus 21, **124**
obsoleta, Amara 23
obsoletus, Dicheirotrichus 25, 167, **168**
obtusiusculus, Tachys 14, 71, **72**
obtusum, Bembidion 18, 97, **99**, 100, 102
obtusus, Trechus 14, 66, **67**, 68, 185
ochraceus, Pterostichus 19
ochropterus, Calathus 20
octomaculatum, Bembidion 17, **99**
OCYDROMUS 16, 80, **87**, 90-93
OCYS 8, 15, 70, **74**
ODACANTHA 2, 3, 29, **199**
Odacanthini 29, 38, 199
ODONTONYX 20, 21
OLISTHOPUS 21, 117, **123**
OMASEUS 19, **113**
OMOPHRON 1, 3, 10, **33**
Omophroninae 10, 33
OODES 27, **179**, 180
Oodini 27, 37, 179
OPHONUS 1, 4, 5, 8, 24, 148, 149, 151, 158, **159**, **161**, 162, 167
OREOCARABUS 11, **42**
orinomum, Pterostichus 19
ovata, Amara 23, 135, 137, 141, **142**
OXYPSELAPHUS 21, 117, **124**

pallens, Plochionus 9, 186
pallidipenne, Bembidion 15, **82**
pallidulus, Tachys 14
pallipes, Asaphidion 15, 73, **74**
pallipes, Paranchus 21

pallorus, Tachys 14
paludosum, Bracteon 15
paludosus, Trechus 14
palustris, Notiophilus 12, 51, **53**
Panagaeini 27, 36, 184
PANAGAEUS 27, **184**
PARACELIA 23, 135, **144**
PARADROMIUS 28, 187, **191**
parallelepipedus, Abax 19, **116**
parallelopipedus, Abax 19
parallelus, Abax 9
parallelus, Ophonus 25, 161-**163**, 164
PARANCHUS 21, 117, **124**
PARANEBRIA 12, **48**
PARATACHYS 14, **71**
PARDILEUS 23
parumpunctatum, Agonum 22
parvulus, Acupalpus 26, 175-**177**
parvulus, Elaphropus 15, **72**
patricia, Amara 23
Patrobini 18, 36, 103
PATROBUS 18, **103**
PEDIUS 19, **111**
pelidnum, Agonum 21
PELOPHILA 12, 45, **50**
peltatus, Badister 27, 182, 183, **184**
PERCOSIA 23, 135, **146**
PERIGONA 27, **185**
Perigonini 7, 27, 36, 185
PERILEPTUS 14, 63, **64**
PERYPHUS 16
PHILA 18
PHILOCHTHUS 17, 78, 80, **100**
PHILOCTHUS 18
PHILORHIZUS 29, 187, 191, 192, **194**
PHYLA 18, 78, **99**
piceum, Agonum 21, 126, **128**
piceus, Calathus 20
piceus, Tachys 14
picimanus, Pterostichus 19
picipennis, Harpalus 24
pilicornis, Loricera 13, **56**
placidus, Trichocellus 26, 168, **169**
plagiatus, Stenolophus 172
planus, Sphodrus 20

PLATAPHUS 16, 80, **84**, 85
PLATYDERES 20
PLATYDERUS 20, 117, **118**
Platynini 7, 21, 37, 116
PLATYNUS 21, 117, **125**
PLATYSMA 19, **112**
plebeia, Amara 22
plebeja, Amara 22, **135**
poeciloides, Anisodactylus 25, 165, **166**
POECILUS 8, 18, 105, **106**, 107, 166
Pogonini 18, 36, 102
POGONOPHORUS 11, **46**
POGONUS 18, **102**
POLISTICHUS 1, 29, 190, **199**
politus, Dyschirius 13, 59, **61**
POLYSTICHUS 29
praetermissa, Amara 23, 135, 143, **144**
praeustus, Leistus 11
prasinum, Bembidion 16, **84**, 85, 87
prasinus, Anchomenus 21
PRINCIDIUM 15, 77, **81**
PRISTONYCHUS 20, **123**
problematicus ssp. *feroensis*, Carabus 11
problematicus ssp. *harcyniae*, Carabus 11
problematicus, Carabus 11, 40-44
properans, Bembidion 15, **81**
pseudoaeneus, Anisodactylus 25
PSEUDOLIMNAEUM 17, **94**
PSEUDOMASEUS 19, **114**
PSEUDOOPHONUS 8, 23, 149, **150**, 158
PSEUDOPHONUS 23
Psydrinae 7
Pterostichini 7, 18, 37, 103, 105, 148
PTEROSTICHUS 4, 7, 8, 18, 19, 105, **107**, 111, 123, 147, 166
pubescens, Dicheirotrichus 25
pubescens, Harpalus 23
puellum, Agonum 21
pumicatus, Stomis 18, **105**
pumilus, Harpalus 24, 149, 151, **156**
punctatulus, Licinus 27, 180, **181**
punctatulus, Ophonus 25
puncticeps, Ophonus 25, 160, 163, **164**, 165
puncticollis, Ophonus 25, 161-**164**
punctulatum, Bembidion 15, **81**

punctulatus, Licinus 27
pusillus, Notiophilus 12
quadrifoveolatus, Pterostichus 19, 109, **113**
quadriguttatum, Bembidion 17
quadriguttatus, Notiophilus 12
quadrillum, Lionychus 26, **197**
quadrimaculatum, Bembidion 17, 97, **98**, 99
quadrimaculatus, Dromius 28, 192, **193**, 194
quadrinotatus, Calodromius 28
quadripunctata, Sericoda 21, **125**
quadripunctatus, Harpalus 24
quadripunctatus, Notiophilus 12, 51-53
quadripustulatum, Bembidion 17, 93, 97, **98**
quadripustulatus, Panagaeus 27
quadrisignatus, Elaphropus 9
quadrisignatus, Philorhizus 29, 193-**195**
quadristriatus, Trechus 14, 66, **67**, 185
quenseli, Amara 23, **144**, 145
quenselii, Amara 23
quinquestriatus, Ocys 15, 74, **75**

rectangulus, Ophonus 25
redtenbacheri, Bembidion 16
rhaeticus, Pterostichus 8, 19, 110, 113, 114, **115**
riparium, Bembidion 18
riparius, Elaphrus 13, 55, **56**
RISOPHILUS 28, **189**
rivularis, Trechus 14, **66**-68
robini, Aepus 14
robinii, Aepus 14, 64, **65**
rotundatus, Olisthopus 21, **124**
rotundatus, Platynini 21
rotundicollis, Calathus 20, 119, **120**
rotundicollis, Ophonus 24
rotundicollis, Platynini 21
rubens, Trechus 14, 65, 67, **68**
rubripes, Harpalus 24, 151, 153-**156**
rufescens, Leistus 11
rufescens, Nebria 12, 48, **49**
rufescens, Ocys 15
rufibarbis, Ophonus 25, 161-**164**, 165
ruficollis, Bradycellus 26, 170, **171**
ruficollis, Platyderus 20
ruficornis, Harpalus 23
ruficornis, Paranchus 21

rufimanus, Harpalus 24
rufipalpis, Harpalus 24, 152, **154-156**
rufipes, Harpalus 23, 150, **151**
rufipes, Notiophilus 12, 51, **53**
rufipes, Patrobus 18
rufitarsis, Harpalus 24
rufocincta, Amara 23
rufomarginatus, Leistus 11, 45, **46**, 47
rupestre, Bembidion 16
rupicola, Ophonus 25, 160, 163, **164**
rupicoloides, Ophonus 25

sabulicola, Ophonus 24, 159, 161, **162**
sabulosa, Nebria 12
salina, Nebria 12, 48, **49**
salinus, Dyschirius 13, 60-**62**
saxatile, Bembidion 16, 87, 90-**92**
scapularis, Lebia 9, 187
Scaritini 3, 13, 35, 57
schaubergerianus, Ophonus 25, 161, 163-**165**
schrankii, Chlaenius 27
schueppeli, Bembidion 17
schuppelii, Bembidion 17, **95**
scitulum, Agonum 21, 127, **128**
sclopeta, Brachinus 10, 32, **33**
scotus, Pterostichus 19
scutellaris, Tachys 14, 71, **72**
SCYBALICUS 25, 148, 149, **167**
secalis, Trechus 14, 66, **67**
seladon, Ophonus 25
SEMICAMPA 17, 78, **94**, 96
semipunctatum, Bembidion 16, **83**, 84, 96
septentrionis, Patrobus 18, 103, **104**
SERICODA 21, 117, **125**
seriepunctatus, Harpalus 24
serripes, Harpalus 24, 153, 155, **157**
servus, Harpalus 24, 153, **157**
sexpunctatum, Agonum 22, 129, 131, **132**
sharpi, Bradycellus 26, 170-**172**
sigma, Philorhizus 29, **194-196**
silphoides, Licinus 27
silvatica, Cicindela 10
silvaticus, Carabus 10
similata, Amara 23, 135, 137, 140, 141, **142**
similis, Bradycellus 26

SINECHOSTICTUS 17, 80, **93**
skrimshiranus, Stenolophus 26, 172, **173**
smaragdinus, Harpalus 24, 152, 155, **157**
sobrinus, Harpalus 24
sodalis, Badister 27, 177, 181, **183**, 184
Sphodrini 7, 20, 37, 116
SPHODRUS 20, 116, **122**
spilotus, Calodromius 28, 193, **194**, 195
spinibarbis, Leistus 11, 46, **47**
spinipes, Curtonotus 23
spreta, Amara 23, 136, 138, 140, 141, **142**
spurcaticornis, Anisodactylus 25
STENOLOPHUS 26, 149, **172**
stephensi, Bembidion 17
stephensii, Bembidion 17, 88, 90, **92**
STEROPUS 19, **111**
stictus, Ophonus 24, 159, 161, **162**
stierlini, Asaphidion 8, 15, 73, **74**
STOMIS 2, 18, **105**
stomoides, Bembidion 17, **93**
strenua, Amara 22, **135**
strenuus, Pterostichus 19, 110, 114, **115**
strenuus, Pterostichus 19
striatulus, Badister 27
strigifrons, Notiophilus 12
striola, Abax 19
sturmi, Bembidion 17
sturmii, Bembidion 17
subcyaneus, Laemostenus 20
subnotatus, Trechus 14, 66-**68**
subpunctatus, Ophonus 25
subquadratus, Ophonus 24
subsinuatus, Ophonus 8, 25, 158, 161
substriatus, Notiophilus 12, 51-**53**
sycophanta, Calosoma 10, **39**
sylvatica, Cicindela 10, 31, **32**
SYNECHOSTICTUS 17
SYNTOMUS 26, 186, 196, **197**, 198
SYNUCHUS 20, 116, **118**, 123

TACHYS 14, 15, 70, **71**, 72
TAPHRIA 20
tardus, Harpalus 24, 153-155, **157**, 158
TARULUS 28, **191**
tenebrioides, Zabrus 22, **133**

tenebrosus ssp. *centralis, Harpalus* 24
tenebrosus, Harpalus 24, **158**
terminatus, Leistus 11, 46, **47**
terricola, Laemostenus 20, **123**
testaceum, Bembidion 17, 88-**92**
TESTEDIUM 15, 77, **82**
tetracolum, Bembidion 17, 89-91, **93**
tetragrammum, Bembidion 17
teutonus, Stenolophus 26, 172, **173**
THALASSOPHILUS 14, 63, **68**
thoracicus, Dyschirius 13, 58, **60**, 61
thoreyi, Agonum 125, 126, 127, **128**
tibiale, Bembidion 16, 84, 86, **87**
tibialis, Amara 23, 137, **142**, 144
Trechidae 7
Trechinae 7
Trechini 2, 13, 35, 63
Trechitae 7
TRECHOBLEMUS 14, 63, **69**
TRECHUS 2, 14, 63, **65**, **67**, 68, 69, 185
TREPANEDORIS 17, 78, **98**
TREPANES 17, 78, **98**, 99
TRIAENA 22
TRICHELAPHRUS 13, **56**
TRICHOCELLUS 25, 148, **168**, 172
TRICHOPLATAPHUS 16, **80**
TRIMORPHUS 27, **183**
tristis, Chlaenius 27, 178, **179**
tristis, Dyschirus 13
trivialis, Amara 22
truncatellus, Syntomus 29, 197, **198**

uliginosus, Elaphrus 13, 55, **56**
unicolor, Bembidion 18
unicolor, Dyschirus 13
unicolor, Lionychus 29
unifasciatus, Somotrichus 9, 186
unipunctatus, Demetrias 28
unipustulatus, Badister 27, 181-**183**
ustulatum, Bembidion 16, 17

vaporariorum, Cymindis 28, 190, **191**
vaporariorum, Stenolophus 26
varium, Bembidion 16, 82-**84**
vectense, Bembidion 16

vectensis, Amara 22
vectensis, Philorhizus 29, 195, **196**
velox, Bembidion 15
verbasci, Bradycellus 26, 170-**172**
vernalis, Harpalus 24
vernalis, Pterostichus 19, 108, 112, **115**
versicolor, Poecilus 18, 106, **107**
versutum, Agonum 22, 130, **132**
vespertinus, Stenolophus 26
vestitus, Chlaenius 27, 178, **179**
viduum, Agonum 22, 130, **133**
violaceus ssp. *purpurascens, Carabus* 11
violaceus ssp. *sollicitans, Carabus* 11
violaceus, Carabus 11, 40, 42-**44**
virens, Bembidion 16, 84, **85**
vitreus, Pterostichus 19
vittatus, Polistichus 29
vivalis, Synuchus 20, **118**
vulgaris, Amara 22
vulgaris, Pterostichus 19

walkerianus, Elaphropus 15, **72**
wetterhali, Masoreus 28
wetterhallii, Masoreus 28, **185**

Zabrini 3-5, 22, 37, 133
ZABRUS 4, 22, **133**, 134
ZEZEA 22, 134, **135**
ziegleri, Stenolophus 26
zigzag, Ophonus 25
Zuphiini 7, 29, 38, 190, 199

Colour plates

Subfamily Cicindelinae

1 *Cicindela campestris* — 12.0-17.0 mm
2 *Cylindera germanica* — 9.0-12.0 mm

Subfamily Brachininae

3 *Brachinus crepitans* — 6.0-9.5 mm

Subfamily Omophroninae

4 *Omophron limbatum* — 5.0-6.5 mm

Subfamily Carabinae
Tribe Carabini

5 *Calosoma inquisitor* — 16.0-22.0 mm
6 *Carabus (Eucarabus) arvensis* — 16.0-20.0 mm
7 *Carabus (Limnocarabus) clatratus* — 22.0-30.0 mm
8 *Carabus (Oreocarabus) glabratus* — 22.0-30.0 mm
9 *Carabus (s.str.) granulatus* — 16.0-23.0 mm

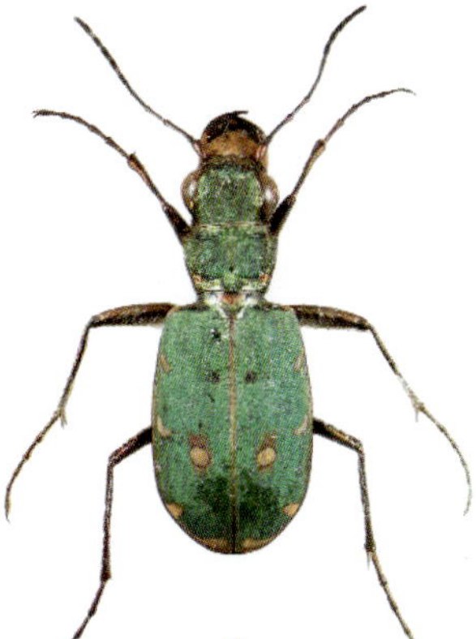

1

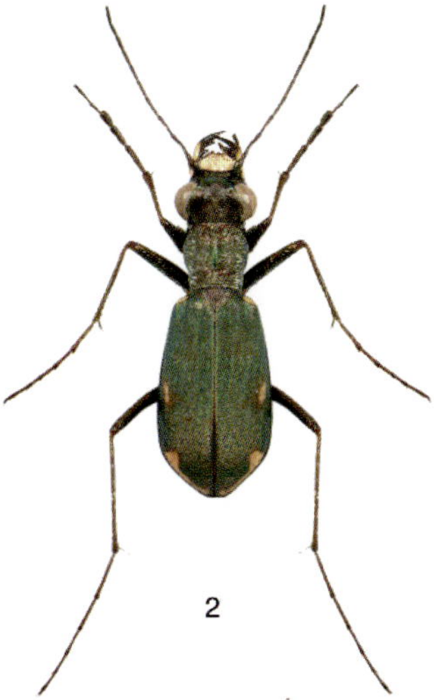

2

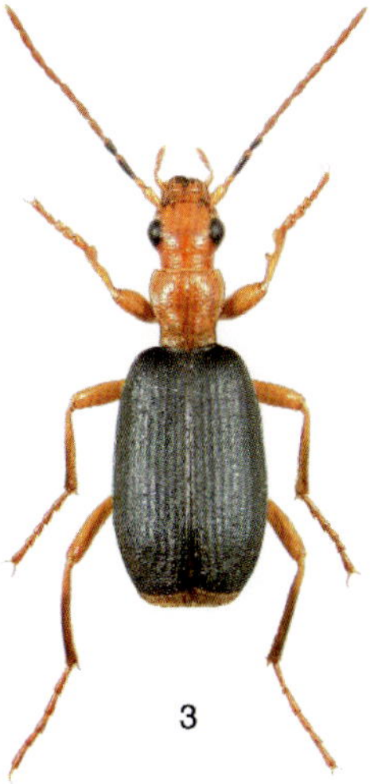

3

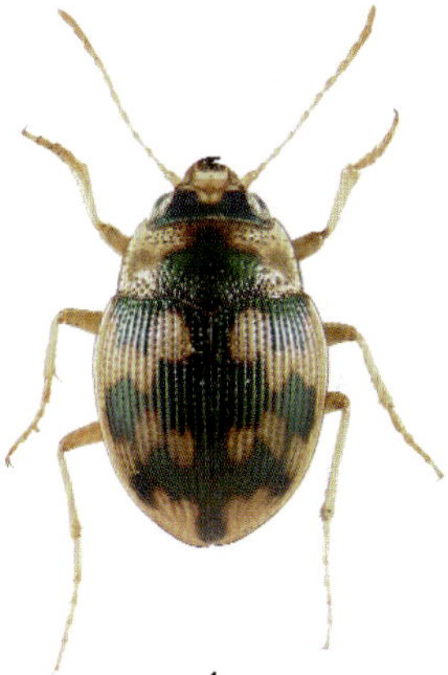

4

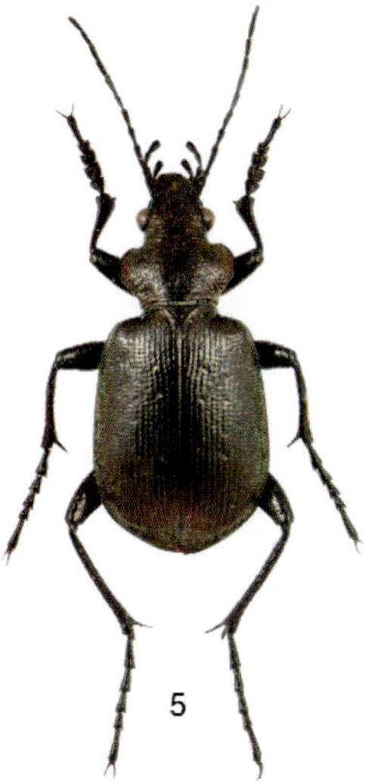

5

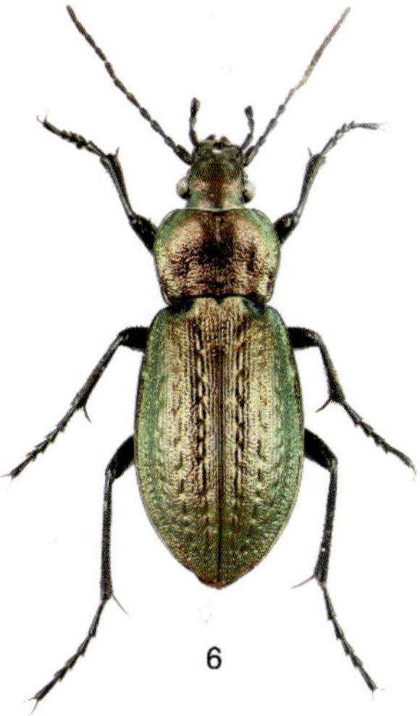

6

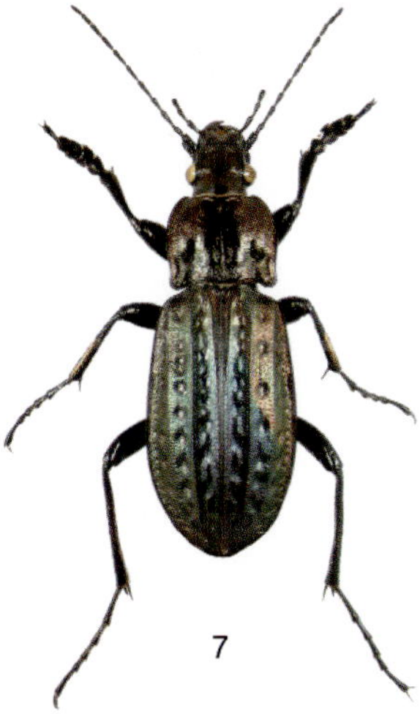

7

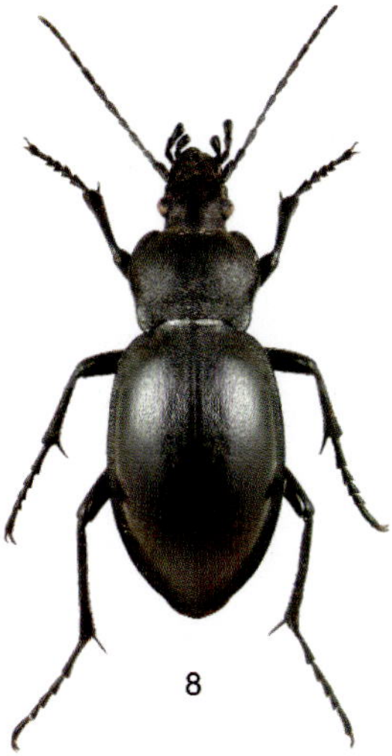

8

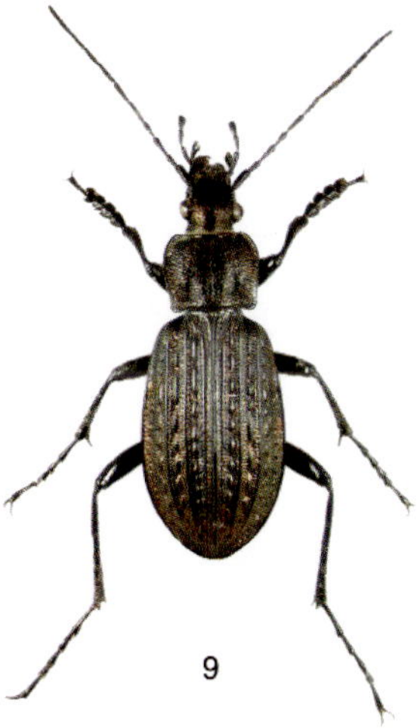

9

Tribe Carabini (continued)

10 *Carabus (Chaetocarabus) intricatus*	25.0-35.0 mm
11 *Carabus (Morphocarabus) monilis*	22.0-26.0 mm
12 *Carabus (Archicarabus) nemoralis*	20.0-26.0 mm
13 *Carabus (Hemicarabus) nitens*	13.0-18.0 mm
14 *Carabus (Mesocarabus) problematicus*	20.0-28.0 mm
15 *Carabus (Megodontus) violaceus*	20.0-30.0 mm

Tribe Cychrini

16 *Cychrus caraboides*	14.0-19.0 mm

Tribe Nebriini

17 *Leistus (Pogonophorus) spinibarbis*	8.0-10.5 mm
18 *Leistus (Leistophorus) fulvibarbis*	6.5-8.5 mm

Handbooks for the Identification of British Insects: Carabidae (ground beetles)

217

Tribe Nebriini (continued)

19 *Leistus (s.str.) ferrugineus*	6.0-8.0 mm
20 *Nebria (Paranebria) livida*	13.0-16.0 mm
21 *Nebria (s.str.) brevicollis*	11.0-14.0 mm
22 *Nebria (Boreonebria) rufescens*	9.0-11.0 mm
23 *Eurynebria complanata*	16.0-23.0 mm
24 *Pelophila borealis*	9.0-12.0 mm

Tribe Notiophilini

25 *Notiophilus biguttatus*	5.0-6.0 mm

Tribe Elaphrini

26 *Blethisa multipunctata*	10.5-13.5 mm
27 *Elaphrus (s.str.) cupreus*	8.0-9.5 mm

19   20   21

22   23   24

25   26   27

Tribe Elaphrini (continued)

28 *Elaphrus (Trichelaphrus) riparius*   6.5-7.5 mm

Tribe Loricerini

29 *Loricera pilicornis*   6.0-8.0 mm

Tribe Scaritini

30 *Clivina fossor*   6.0-6.8 mm

31 *Dyschirius (s.str) thoracicus*   3.5-4.5 mm

32 *Dyschirius (Dyschiriodes) globosus*   2.3-2.9 mm

Tribe Broscini

33 *Broscus cephalotes*   17.0-22.0 mm

34 *Miscodera arctica*   6.5-8.0 mm

Tribe Trechini

35 *Perileptus areolatus*   2.3-2.8 mm

36 *Aepus robinii*   2.2-2.9 mm

Handbooks for the Identification of British Insects: Carabidae (ground beetles)

28 29 30
31 32 33
34 35 36

Tribe Trechini (continued)

37	*Trechus (Epaphius) secalis*	3.5-4.0 mm
38	*Trechus (s.str.) quadristriatus*	3.6-4.1 mm
39	*Thalassophilus longicornis*	3.5-4.0 mm
40	*Blemus discus*	4.5-5.5 mm
41	*Trechoblemus micros*	3.8-4.5 mm

Tribe Bembidiini

42	*Tachys (Paratachys) micros*	2.0-2.4 mm
43	*Tachys (s.str.) scutellaris*	2.0-2.6 mm
44	*Elaphropus parvulus*	1.8-2.2 mm
45	*Asaphidion flavipes*	3.9-4.7 mm

37 38 39

40 41 42

43 44 45

Tribe Bembidiini (continued)

46	*Ocys harpaloides*	4.2-5.8 mm
47	*Cillenus lateralis*	3.4-4.2 mm
48	*Bracteon litorale*	5.6-6.2 mm
49	*Bembidion (Neja) nigricorne*	3.3-4.0 mm
50	*Bembidion (Metallina) lampros*	3.0-4.0 mm
51	*Bembidion (Princidium) punctulatum*	4.7-5.7 mm
52	*Bembidion (Actedium) pallidipenne*	4.3-4.9 mm
53	*Bembidion (Testedium) bipunctatum*	3.9-4.7 mm
54	*Bembidion (Eupetedromus) dentellum*	5.2-6.0 mm

Handbooks for the Identification of British Insects: Carabidae (ground beetles)

46 47 48

49 50 51

52 53 54

Tribe Bembidiini (continued)

55	*Bembidion (Notaphus) varium*	4.1-5.0 mm
56	*Bembidion (Notaphemphanes) ephippium*	2.5-3.6 mm
57	*Bembidion (Plataphus) prasinum*	4.5-5.4 mm
58	*Bembidion (Trichoplataphus) virens*	4.4-5.2 mm
59	*Bembidion (Bembidionetolitzkya) tibiale*	5.5-6.5 mm
60	*Bembidion (Ocydromus) tetracolum*	5.0-6.0 mm
61	*Bembidion (Nepha) illigeri*	4.1-5.2 mm
62	*Bembidion (Sinechostictus) stomoides*	5.2-6.2 mm
63	*Bembidion (Lymnaeum) nigropiceum*	3.6-4.1 mm

55 56 57

58 59 60

61 62 63

Tribe Bembidiini (continued)

64 *Bembidion (Semicampa) schuppelii*	2.9-3.4 mm
65 *Bembidion (Diplocampa) fumigatum*	3.4-3.9 mm
66 *Bembidion (Emphanes) minimum*	2.4-3.3 mm
67 *Bembidion (s.str.) quadrimaculatum*	2.8-3.4 mm
68 *Bembidion (Trepanodoris) doris*	3.1-3.7 mm
69 *Bembidion (Trepanes) articulatum*	2.9-3.9 mm
70 *Bembidion (Phyla) obtusum*	2.8-3.5 mm
71 *Bembidion (Philochthus) biguttatum*	3.9-4.4 mm

Tribe Pogonini

72 *Pogonus chalceus*	5.5-7.0 mm

Handbooks for the Identification of British Insects: Carabidae (ground beetles)

64 65 66

67 68 69

70 71 72

Tribe Patrobini

73 *Patrobus atrorufus* 7.5-9.5 mm

Tribe Pterostichini

74 *Stomis pumicatus* 6.5-8.5 mm
75 *Poecilus cupreus* 11.0-13.0 mm
76 *Pterostichus (s.str.) cristatus* 14.0-17.0 mm
77 *Pterostichus (Steropus) madidus* 14.0-18.0 mm
78 *Pterostichus (Pedius) longicollis* 5.0-6.0 mm
79 *Pterostichus (Lyperosomus) aterrimus* 13.0-15.0 mm
80 *Pterostichus (Adelosia) macer* 10.5-14.0 mm
81 *Pterostichus (Playtsma) niger* 16.0-21.0 mm

Handbooks for the Identification of British Insects: Carabidae (ground beetles)

231

Tribe Pterostichini (continued)

82 *Pterostichus (Bothriopterus) adstrictus*	10.0-13.0 mm
83 *Pterostichus (Omaseus) melanarius*	13.0-17.0 mm
84 *Pterostichus (Pseudomaseus) nigrita*	9.0-12.0 mm
85 *Pterostichus (Lagarus) vernalis*	6.0-7.5 mm
86 *Pterostichus (Argutor) strenuus*	6.0-7.2 mm
87 *Abax parallelepipedus*	17.0-22.0 mm

Tribe Sphodrini

88 *Platyderus depressus*	6.0-8.5 mm
89 *Synuchus vivalis*	6.0-8.5 mm
90 *Calathus (Amphigynus) rotundicollis*	8.5-10.5 mm

Handbooks for the Identification of British Insects: Carabidae (ground beetles)

82

83

84

85

86

87

88

89

90

Tribe Sphodrini (continued)

91 *Calathus (s.str.) fuscipes*	10.0-14.0 mm
92 *Sphodrus leucophthalmus*	21.0-28.0 mm
93 *Laemostenus (Pristonychus) terricola*	13.0-17.0 mm

Tribe Platynini

94 *Olisthopus rotundatus*	6.5-8.0 mm
95 *Oxypselaphus obscurus*	5.0-6.5 mm
96 *Paranchus albipes*	6.5-8.8 mm
97 *Anchomenus dorsalis*	6.0-8.0 mm
98 *Platynus assimilis*	9.0-12.5 mm
99 *Batenus livens*	7.5-10.0 mm

91 92 93

94 95 96

97 98 99

Tribe Platynini (continued)

100	*Agonum (Europhilus) fuliginosum*	5.5-7.0 mm
101	*Agonum (s.str.) muelleri*	7.0-9.0 mm

Tribe Zabrini

102	*Zabrus tenebrioides*	14.0-16.0 mm
103	*Amara (Zezea) plebeja*	6.0-7.8 mm
104	*Amara (s.str.) aenea*	6.5-8.8 mm
105	*Amara (Celia) bifrons*	5.5-7.3 mm
106	*Amara (Paracelia) quenseli*	6.5-9.5 mm
107	*Amara (Bradytus) apricaria*	6.5-8.5 mm
108	*Amara (Percosia) equestris*	7.5-10.5 mm

Handbooks for the Identification of British Insects: Carabidae (ground beetles)

100

101

102

103

104

105

106

107

108

Tribe Zabrini (continued)

109 *Curtonotus aulicus*   11.0-14.0 mm

Tribe Harpalini

110 *Harpalus (Pseudoophonus) rufipes*   11.0-16.0 mm
111 *Harpalus (s.str.) affinis*   9.0-12.0 mm
112 *Harpalus (Cryptophonus) tenebrosus* 8.0-11.0 mm
113 *Ophonus (s.str.) ardosiacus*   9.5-11.5 mm
114 *Ophonus (Metophonus) puncticeps*   6.5-9.0 mm
115 *Anisodactylus binotatus*   10.0-13.0 mm
116 *Dicheirotrichus gustavii*   5.5-7.5 mm
117 *Trichocellus placidus*   4.0-5.5 mm

Handbooks for the Identification of British Insects: Carabidae (ground beetles)

109

110

111

112

113

114

115

116

117

Tribe Harpalini (continued)

118	*Bradycellus harpalinus*	3.8-5.0 mm
119	*Stenolophus mixtus*	5.0-6.0 mm
120	*Acupalpus dubius*	2.5-2.8 mm
121	*Anthracus consputus*	3.8-5.0 mm

Tribe Chlaeniini

122	*Chlaenius nigricornis*	10.0-12.0 mm
123	*Callistus lunatus*	6.0-7.0 mm

Tribe Oodini

124	*Oodes helopioides*	7.5-10.0 mm

Tribe Licinini

125	*Licinus depressus*	9.0-11.5 mm
126	*Badister (s.str.) bullatus*	4.8-6.3 mm

118 119 120

121 122 123

124 125 126

Tribe Licinini (continued)

127 *Badister (Trimorphus) sodalis*	3.8-4.6 mm
128 *Badister (Baudia) dilatatus*	4.5-5.4 mm

Tribe Panagaeini

129 *Panagaeus cruxmajor*	7.4-8.8 mm

Tribe Perigonini

130 *Perigona nigriceps*	2.0-2.5 mm

Tribe Masoreini

131 *Masoreus wetterhallii*	4.5-6.0 mm

Tribe Lebiini

132 *Lebia (Lamprias) chlorocephala*	5.8-8.0 mm
133 *Lebia (s.str.) cruxminor*	5.0-7.0 mm
134 *Demetrias (Risophilus) imperialis*	4.5-5.8 mm
135 *Demetrias (s.str.) atricapillus*	4.5-5.5 mm

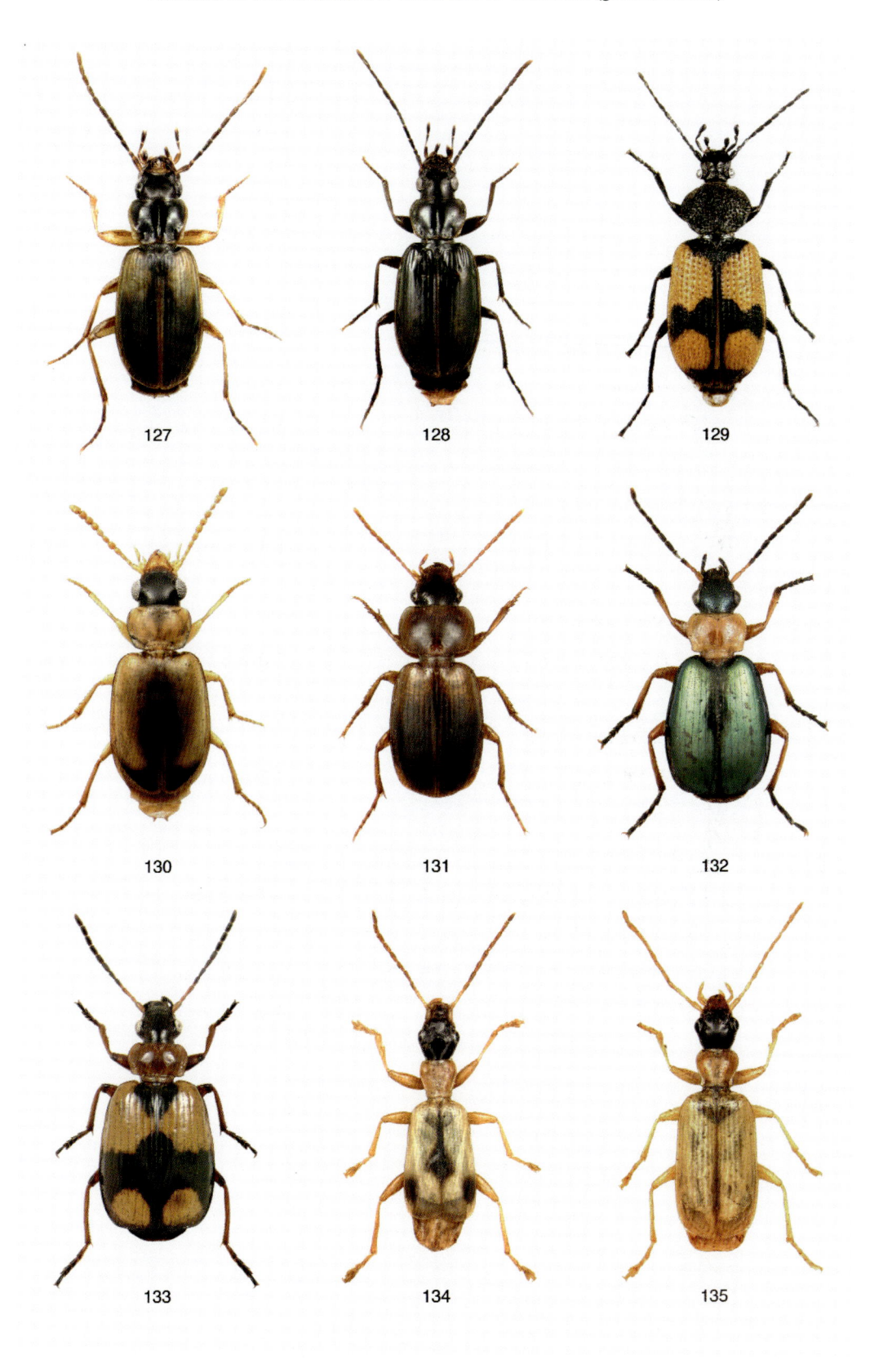

Handbooks for the Identification of British Insects: Carabidae (ground beetles)

127 128 129

130 131 132

133 134 135

243

Tribe Lebiini (continued)

136 *Cymindis (s.str.) axillaris*	8.0-10.5 mm
137 *Cymindis (Tarulus) vaporariorum*	8.0-10.0 mm
138 *Paradromius linearis*	4.5-5.5 mm
139 *Dromius quadrimaculatus*	5.0-6.0 mm
140 *Calodromius spilotus*	3.8-4.5 mm
141 *Philorhizus melanocephalus*	2.8-3.5 mm

136

137

138

139

140

141

245

Tribe Lebiini (continued)

142 *Microlestes minutulus*	2.8-3.7 mm
143 *Lionychus quadrillum*	3.0-4.0 mm
144 *Syntomus foveatus*	3.2-3.8 mm

Tribe Odacanthini

145 *Odacantha melanura*	6.5-7.5 mm

Tribe Dryptini

146 *Drypta dentata*	7.0-8.5 mm

Tribe Zuphiini

147 *Polistichus connexus*	8.0-9.5 mm

142 143 144

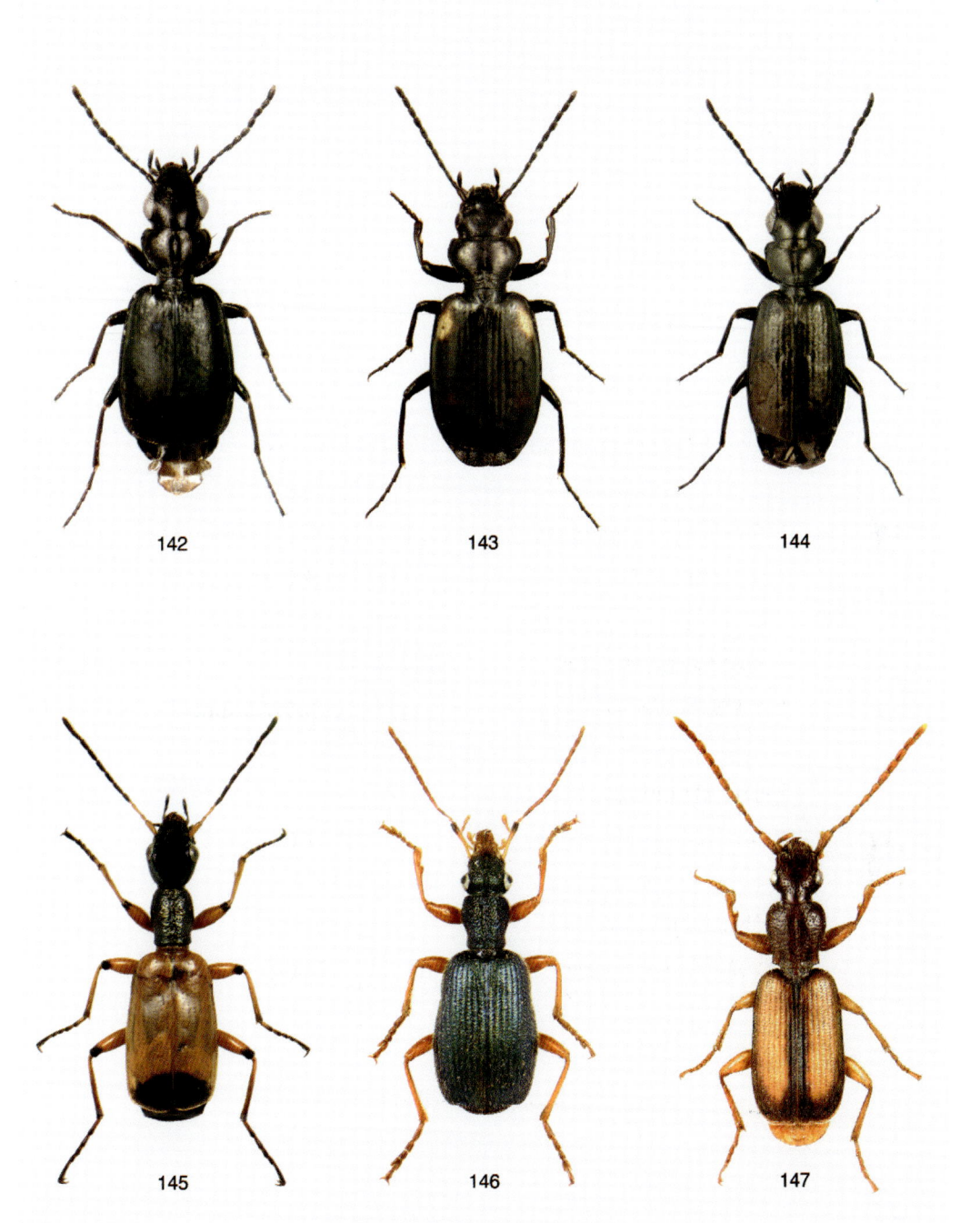

145 146 147